Creating Responsible Inquiring Digital Health Organizations

Nilmini Wickramasinghe
La Trobe University, Australia

Nalika Ulapane
La Trobe University, Australia

Ayesha Nilashini
La Trobe University, Australia

IGI Global
Scientific Publishing
Publishing Tomorrow's Research Today

Published in the United States of America by
IGI Global Scientific Publishing
701 East Chocolate Avenue
Hershey, PA, 17033, USA
Tel: 717-533-8845 | Fax: 717-533-7115
Website: https://www.igi-global.com E-mail: cust@igi-global.com

Library of Congress Cataloging-in-Publication Data

Names: Wickramasinghe, Nilmini author | Ulapane, Nalika, 1988- author | Nilashini, Ayesha, 1987- author

Title: Creating responsible inquiring digital health organizations / edited by Nilmini Wickramasinghe, Nalika Ulapane, Ayesha Nilashini.

Description: Hershey, PA : IGI Global Scientific Publishing, [2026] | Includes bibliographical references and index. | Summary: "Healthcare is facing many challenges including escalating costs, workforce challenges and patients demanding more personalization. Contemporaneously there is the rapid rise of technology especially artificial intelligence and more specifically Generative AI. Thus to leverage these technology advances it is essential to create responsible inquiring digital health organizations"-- Provided by publisher.f

Identifiers: LCCN 2025024364 (print) | LCCN 2025024365 (ebook) | ISBN 9798337318776 hardcover | ISBN 9798337318783 paperback | ISBN 9798337318790 ebook

Subjects: LCSH: Artificial intelligence--Medical applications | Medical informatics | Knowledge management

Classification: LCC R859.7.A78 W53 2026 (print) | LCC R859.7.A78 (ebook)

LC record available at https://lccn.loc.gov/2025024364

LC ebook record available at https://lccn.loc.gov/2025024365

British Cataloguing in Publication Data

A Cataloguing in Publication record for this book is available from the British Library.

For our Families, Colleagues, Readers and All Who Research in Healthcare

Table of Contents

The Process Perspective

Foreword

We are privileged to be living in a period of extraordinary, exponential advances in computing power, software applications, artificial intelligence (AI), and the availability and insightful analysis of massive volumes of real world data – exciting advances that collectively are creating powerful opportunities for the transformation of how critical knowledge is intelligently developed, aggregated, and deployed.

However, doing so demands more than simply massive developments in computing power, AI, machine learning, agentic tools and other technological advances, it also requires thoughtful knowledge creation, ethical and sophisticated knowledge management, trustworthy governance, and sociotechnical considerations that together support the realities of clinical care and deliver the health benefits desired by all patients.

For healthcare providers, both clinicians and healthcare organisations, AI-enabled tools can narrow diagnostic uncertainty, personalise treatment, predict deterioration, and streamline administration and workflow. For receivers of care, these developments translate into agency, transparency, and better outcomes, with the potential for care to become more coordinated, timely, and person-centred. For funders and regulators, AI-interpreted data and knowledge offer potential for the ubiquitous adoption of true value-based care.

The promise of modern computing, and especially AI and machine learning applied in digital health, can only be fully realised when diverse, heterogeneous, relevant data is converted into reliable, context-aware knowledge that is developed and deployed responsibly, transparently and safely – key principles that apply equally to the very nature, standards and ethics of healthcare delivery itself.

In this book, *Creating Responsible Intelligent Digital Health Organisations*, edited by Professor Nilmini Wickramasinghe, Dr Nalika Ulapane and Ayesha Nilashini, the editors have set out both contemporary as well as forward looking chapters exploring key themes of knowledge management, advances and application of AI and machine learning in healthcare and, crucially, the application and utility of these advances in specific healthcare settings.

This book is a truly valuable resource for all readers who have an interest in understanding the combination of the rapid evolution of AI, machine learning and related computing technologies, their role in knowledge generation and management, and specifically the potential for significant advances in the quality and effectiveness of healthcare in high value and key application areas.

It is indeed relevant and timely. I recommend it highly.

John Zelcer
La Trobe University, Australia

Preface

INTRODUCTION

Today healthcare delivery in OECD countries is facing many challenges post COVID. These include escalating healthcare cots to treat everyone with the latest technology and medical advances, a rapid rise of chronic conditions, an aging population and life expectancy, a work force shortage and an existing workforce that is jaded and time poor.

Taken together this has severe impacts on the quality of care, level of personalization and value of the care received. It is not possible to solve these challenges using the same approaches. We content that it is only possible to make a substantive impact by leveraging the sizeable advances in technology, especially artificial intelligence (AI) and machine learning (ML). However, we believe it will only be truly possible to realize the full potential of AI and ML if we take a knowledge management approach and frame healthcare organizations as intelligent inquiring organizations.

The following thus presents the rationale for "Creating Responsible Intelligent Digital Health Organizations." Specifically, we discuss in the following 12 chapters the critical issues that must be considered. This is done in three sections taking a socio-technical approach. Section I primarily presents the people perspective. It consists of six chapters as follows

ORGANIZATION OF THE BOOK

Section I: The People Perspective

Chapter 1: Overview of the Networked Knowledge Environment

This chapter presents the critical aspects of the knowledge environment explaining how raw data can be transformed into meaning full and important knowledge assets and how this knowledge forms the key in a knowledge economy for growth and sustainability.

Chapter 2: The Construct of Knowledge

This chapter unpacks the complex and heterogeneous construct of knowledge and identifies its many aspects and why it is so essential in healthcare contexts today.

Chapter 3: Creating Knowledge and Decision Making

As explained in this chapter, knowledge creation is essential for decision making and without prudent knowledge creation or generation sound decisions in healthcare cannot be made successfully.

Chapter 4: Managing Knowledge Complexity

By its very nature, healthcare operations are complex. Managing the complexity of knowledge generated in healthcare operations successful is vital in order to ensure prudent decisions are made and timely care is provided. This chapter uncovers the dynamics surrounding knowledge complexity and how best to navigate this.

Chapter 5: The Organization's Human Infrastructure

Healthcare is a data and knowledge rich industry. It human infrastructure plays a significant role in daily activities which support patients and their care and lead to treatment and clinical outcomes. This chapter enables a deeper understanding of the importance of understanding the role of knowledge workers and recognizing clinicians as such knowledge workers.

Section II focuses on the critical AI technical aspects in three key chapters looking at digital twins, generative AI and agentic AI as follows:

Section II: The Technology Perspective

Chapter 6: Digital Twins in Healthcare Discovering Complex Biological Rules for Achieving Precision and Personalization in Healthcare

Digital twins or virtual replication of physical entities are not a new construct, they have been around for several decades. However, their application into healthcare contexts is new. This chapter introduces the importance and potential of digital twins to enable more precise and personalized decision making to ensue and illustrates this in the context of cancer.

Chapter 7: Generative AI Opportunities in Healthcare

This chapter serves to introduce the possibilities and potential of generative AI for various healthcare scenarios. It emphasizes the strengths of such a perspective but also presents some limitations and risks that could occur if a responsible approach is not adopted.

Chapter 8: Agentic AI Opportunities In Healthcare

This chapter serves to introduce the possibilities and potential of agentic AI for various healthcare scenarios. It emphasizes the strengths of such a perspective but also presents some limitations and risks that could occur if a responsible approach is not adopted.

The last section, Section III focuses on the process perspective. This section consists of four chapters covering the important topics of Chapter 9: Strategic Knowledge Management, which serves to highlight the importance of employing various presented strategic tactics and tools to ensure optimization of any and all processes. Chapter 10: Preventative Care, which illustrates how combining KM tools and techniques to best manage wellness and a health lifestyle can prevent critical health problems such as diabetes and thus enable an individual to enjoy a high quality of life and reduce healthcare cots and burdens. Chapter 11: Acute Care, which illustrates how the incorporation of KM tools and techniques and technologies can reduce critical healthcare incidents and speed up recovery and return to a normal or healthier state. And Chapter 12: Rehab and Aged Care, which illustrates how the incorporation of specific KM tools and techniques can ensure healthier and happier aging be it at home or in a care facility. Finally, the Epilogue brings things together and underscores the key benefits and take aways from understanding the opportunities and benefits of technological advances such as AI and ML in terms of and through a KM lens.

CONCLUSION

Clearly, in one volume it is not possible to cover in depth all essential areas. Rather the book is structured to introduce the reader to the key topics and provide a foundation and bedrock. From here, the reader can then move forward with confidence having a solid grounding in core concepts and fundamental principle so will be well able to navigate the road ahead and ensure that the advances in technology especially with respect to Ai and machine learning can be suitably leveraged to provide and/ or support superior healthcare operations. We trust our readers enjoy the material in the following pages at least as much as we have enjoyed compiling the material.

Nilmini Wickramasinghe
La Trobe University, Australia

Nalika Ulapane
La Trobe University, Australia

Ayesha Nilashini
La Trobe University, Australia

Acknowledgement

This book would not have been possible without the efforts, support and encouragement of so many people, too numerous to mention. We wish to thank our institution, La Trobe University, in particular our School of Computing, Engineering and Mathematical Sciences and all our colleagues all over the world. Further, we thank our families and friends. In addition, we have greatly appreciated and valued the support the team at IGI provided. Finally, we thank you the reader and hope the following pages provide you with assistance and inspiration.

Section 1
The People Perspective

Chapter 1
Overview of the Networked Knowledge Environment

ABSTRACT

Today we find ourselves in the knowledge economy or epoch. Previous epochs included the agrarian in which land and labour were essential and farming contributed the most to GDP while in contrast in the industrial epoch land labour and capital were essential and manufacturing contributed most to GDP. In the knowledge epoch data and information are vital and turning these data and information into knowledge assets is what contributes most to an increase in GDP. Given that healthcare is so data rich and extracting actionable knowledge and pertinent information is essential to support clinical decision making it behoves us to understand the knowledge environment and its unique qualities and nuances as is presented in this chapter.

INTRODUCTION

This chapter introduces the fundamental concept of NKE, inquiring organizations, motivations for KM (knowledge management), and how organizations can move toward becoming knowledge-based businesses. The objective is to offer an understanding of how the company may improve organizational productivity and competitiveness within the context of a constantly changing digital environment.

DOI: 10.4018/979-8-3373-1877-6.ch001

KEY CHARACTERISTICS OF NETWORKED KNOWLEDGE ENVIRONMENTS

Collaborative Platforms

Collaborative platforms are the core of NKE as they provide an environment where people and groups can share knowledge and cooperate (Davenport & Prusak, 1998; Välimaa et al., 2016). Apps like Microsoft Teams, Slack, and Trello enhance real-time communication and project management in the working environment (Davenport & Prusak, 1998). These are not mere communication tools, but integrated systems for sharing files and ideas, organizing projects, and data management, which are crucial in a networked environment. Such platforms offer a myriad of benefits to organizations (Välimaa et al. 2016; Davenport & Prusak, 1998) such as:

- Breaking Silos: Open communication platforms eliminate hierarchal barriers within an organization since people from different departments can work together. Workers across various functions perform tasks in teams, engage in knowledge sharing, and address challenges without the rigidity of organizational layers (Brynjolfsson & McAfee, 2014).
- Improved Knowledge Sharing: Slack offers an opportunity to create dedicated channels for specific topics or tasks, which ensures that the required knowledge is always in front of a particular employee. For instance, a company in the process of engaging in sustainability projects can have separate channels for research, planning, and updates on the project.
- Global Collaboration: Communication tools overcome spatial divides, allowing teams to be on different continents but still be efficient. This has become especially crucial in the current world that has embraced working from home as a result of the COVID-19 pandemic (Grant, 2006; Loureiro Cardoso et al, 2024). It becomes apparent why organizations with teams that are spread across different locations depend on such platforms to ensure that everyone is on the same page and working towards the same goals.

Integration with Other Tools: Today's platforms are compatible with productivity apps such as Google Workspace, Microsoft 365, and CRM, making them an all-in-one interface for workers. This integration helps to minimize disruptions and time spent on shifting from one application to another (Al-Qahtani & Aksoy, 2022).

Thus, this increase in collaborative platforms shows that connectiveness and access to information is pivotal in today's economy. These tools bring the much-needed transparency and teamwork to help organizations harness their collective knowledge where necessary.

Decentralization

Another feature of NKE that can be highlighted is decentralization (Grant, 2006). As opposed to conventional models where communication and dissemination of information are centralised at the organisational apex, NKE encourages decentralised utilisation and sharing of knowledge across the entire company (Grant, 2006). This is achieved through (Grant, 2006; Hoadley & Kali, 2019; Loureiro Cardoso et al., 2024);

- Empowered Decision-Making: Mobile knowledge repositories provide decision support, freeing lower-level employees from having to refer to their superiors for information. For instance, a sales representative with immediate access to customer information can adjust their strategy on the fly, enhancing the clients' experience and conversion percentages.
- Faster Problem Solving: This means that when knowledge is available in the public domain, then issues can be solved faster. More decisions can be made at lower levels when everyone has the same information and tools, promoting anticipatory problem-solving (Brynjolfsson & McAfee, 2014).
- Promoting Agility: As the business environment constantly evolves, flexibility is vital in order to remain relevant. Decentralization helps to eliminate bureaucratic barriers to quick adjustments to market conditions. For instance, organizations such as Amazon and Tesla rely on decentralized structures that enable subgroups to work on different projects and implement strategies on their own.
- Trust and Accountability: Knowledge decentralisation promotes trust among employees because it provides them with freedom and information. This sense of trust fosters accountability as the members develop a clear understanding of the expectations within the workplace (Davenport & Prusak, 1998).

Hence, with decision-making power dispersed and knowledge shared across the organization, decentralization is not only beneficial to flexibility but also promotes teamwork and accountability.

Real-Time Access

Real-time data processing is one of the key tenets of the NKE. Thanks to cloud computing, AI, and IoT, real-time access means that employees receive timely information to make the right decisions and improve organizational performance (Brynjolfsson & McAfee, 2014; Hoadley & Kali, 2019). This real-time access is realized through various approaches such as:

- Cloud Computing: Today, organizations are turning to cloud solutions such as Google Cloud and AWS to help them store and process data in real-time from any part of the globe. This is especially valuable for organizations with branches in different countries that need to collaborate on a daily basis.
- Artificial Intelligence: There are also technologies such as chatbots and decision support systems that incorporate AI to facilitate real-time decisions. For instance, a chatbot as a part of the implementation in an e-commerce platform can respond to customer inquiries instantly, enhancing the customers' experience, and alleviating the burden on support staff.
- IoT Applications: IoT devices continuously collect and produce a large volume of real-time data that can help organizations manage operations effectively. For instance, in the manufacturing industry, IoT sensors may identify failed equipment before they halt production (Grant, 2006).
- Data-Driven Decisions: Real-time data enables organizations to shift their decision-making strategy from a reactive one to a proactive one. For example, in a crisis, the availability of up-to-date information empowers the leadership teams to evaluate the situation and allocate resources (Davenport & Ronanki, 2018).

Thus, through providing organizations with the opportunity to receive up-to-date information in real time, NKE helps organizations manage their operations effectively and deal with problems swiftly and accurately.

ADVANTAGES OF NETWORKED KNOWLEDGE ENVIRONMENTS

Enhanced Innovation

One of the most important factors to support innovation is to create conditions that foster collaboration, sharing, and openness (Hoadley & Kali, 2019). It is important to note that NKE offers the necessary tools and frameworks to cultivate such a culture (ibid). In this, NKE does not have a culture that is compartmentalized in terms of departments which means cross functional teams from various departments can work together to develop solutions. For instance, in car manufacturing, the design, engineering, and marketing departments work together to create automobiles that will suit the customers' needs as well as fit the market requirements (Al-Qahtani & Aksoy, 2022). Therefore, ideas can be shared and reviewed by others in a short time thereby increasing the rate of innovation. Many companies like Google and 3M provide opportunities for employees to try out new ideas, which results in innovations like Gmail and Post-it Notes (Grant, 2006). By creating a knowledge repository, NKE allows organizations to quickly recall solutions that have been implemented before, thereby saving time and energy when trying to address similar issues (Hoadley & Kali, 2019). This not only makes work more productive but also helps to develop a culture of constant improvement (Davenport & Prusak, 1998). NKE also promotes open innovation by encouraging organizations to engage with external participants such as suppliers, customers and academic institutions. For instance, it is common for the pharmaceutical industries to collaborate with universities to speed up the process of finding new drugs (Brynjolfsson & McAfee, 2014).

Increased Efficiency

Another advantage of NKE is that it is efficient in its operations and production of products and services (Hoadley & Kali, 2019). Some of the advantages of BPA (business process automation) include the ability to save time and money through the elimination of repetitive and time-consuming activities (BOOK, 2019). AI and machine learning tools minimize the time taken on repetitive tasks like data entry, report preparation, among others. For example, robotic process automation (RPA) is applied in finance and accounting to manage invoices and reconciliation (Davenport & Ronanki, 2018). Real-time data also allows organizations to adjust resources allocation effectively. For example, real-time tracking systems are employed by logistics companies to determine the shortest and most efficient routes to deliver goods to their destinations, thereby minimizing fuel expenses (Al-Qahtani & Aksoy, 2022). This means that the company frees up its human resources through eliminating unnecessary activities so that they can work on activities that need innovative thinking. Moreover, it not only increases efficiency but also enhances the morale and loyalty of employees (Davenport & Prusak, 1998).

NKE also offers great convenience to organizations in that it allows them to expand operations easily without a proportional increase in infrastructure (BOOK, 2019). For instance, cloud-based solutions provide the flexibility to augment or decrease resources as needed, which impacts cost positively. Because knowledge management systems are efficient in terms of the distribution of information and the reduction of unnecessary printing and energy consumption, KM can be considered an environmentally friendly approach. Sustainability-focused organizations can use NKE to support their green endeavors, as the company possesses the necessary experience and resources (Brynjolfsson & McAfee, 2014).

Through the use of transverse platforms, decentralisation and real-time connection, organisations working in a context of networked knowledge can increase their innovations and productivity thus increasing their long-term viability in a competitive global economy.

SOME EXAMPLES OF KNOWLEDGE ENVIRONMENTS (KES)

KEs are crucial particularly to organizations that want to sustain themselves in the global economy which is highly dynamic in the current business world. Alphabet Inc and Apple Inc are good examples of how the NKE principles can be harnessed in building the strong culture of collaboration, innovation, and learning, thus attaining the sustainable competitive advantage. These examples show how organizations systematically coordinate tools, processes, and cultural norms to foster knowledge sharing, improve decisions, and foster innovation.

Google: A Model of Knowledge Sharing and Innovation

The knowledge environment of Google is a prime example of how an organization can benefit from a proper setup. In terms of corporate culture and operation model, it maintains a sustainable innovation environment to ensure efficient knowledge management through (Hoadley & Kali, 2019):

- Collaborative Tools: Google's internal applications, including Google Workspace, facilitate internal collaboration among the company's employees. With the help of collaborative tools such as Google Docs and Google Meet, the company can share information and ideas in real-time, and make sure that all of its employees are on the same page (Schein, 2010).
- Decentralized Decision-Making: Google's organization structure is highly decentralized which means that most of the decision-making powers are devolved down to employee levels. For example, its '20% time policy' lets employees spend a fifth of their working days on initiatives they find inspiring. This freedom fosters innovation and has led to developments such as Gmail or Google Maps, as Von Krogh, Wei, and Makri argue (2012).
- Knowledge Sharing Through Documentation: Google has a well-stocked and easily accessible collection of articles, papers, and case studies for its employees. This knowledge base enables learning and expedites project development, relying on past findings instead of starting from scratch.
- Fostering a Learning Culture: Google also fosters learning by providing opportunities such as internal training programs, workshops, and knowledge-sharing sessions. This is because the employees are supported in developing new knowledge and cross-training practices promote both individual and organizational growth (Schein, 2010).

Through adopting such principles in its operations, Google has remained among the most innovative firms globally, which supports the importance of a coherent knowledge environment.

Apple: Innovating Through Knowledge Integration

This example demonstrates that knowledge management and innovation at Apple are interconnected processes that depend on the integration of design, technology, and business knowledge within the company. Apple has a well-established approach to the development of a new product that involves the design team, engineers, and marketers (Schein, 2010). For instance, the creation of the iPhone involved the convergence of software and hardware engineering as well as user interface and design. This level of collaboration is facilitated by knowledge-sharing systems (Schein, 2010). The leaders at Apple are responsible for establishing and maintaining the knowledge environment. During the reign of Steve Jobs and present CEO Tim Cook, Apple has continued to foster an environment of openness that encourages creativity and questioning of structures (ibid). Leadership makes employees feel appreciated and willing to share their knowledge in projects (von Krogh et al., 2012). Apple also factors in the knowledge environment by considering the customers' feedback and market demands in order to meet the users' needs. This feedback loop is made possible by the use of analytics and data collection in real-time, which shapes product development and marketing (Schein, 2010). By adopting such strategies, Apple continues to provide lead examples of how knowledge environments can be used to fuel innovation in order to maintain a competitive edge in highly saturated markets.

The Broader Impact of Exemplary Knowledge Environments

The cases of Google and Apple illustrate that the creation of a strong knowledge environment is not only about implementing sophisticated technologies, but also about the development of cooperation and learning culture. Organizations in the social sciences are keen to ensure that knowledge-sharing practices are aligned to its goals and objectives so that the employees can support innovation and productivity (Hoadley & Kali, 2019). The effects of such environments are not limited to individual organizations or firms. Indeed, sharing knowledge within the specific settings seen at Google and Apple fosters openness, participation, and preservation of the environment, leading to improvement in other organizations (von Krogh et al., 2012).

ORGANIZATIONS AND KNOWLEDGE: INQUIRY

Inquiring organizations are the ones that are constantly learning and trying to understand their environment better to enhance their performance, generate new ideas, and find solutions to challenges (Lin et al., 2015). Such organizations are marked by their dedication to learning and development, knowledge optimization, and the use of technology. The organisations that use inquiry system have the competitive advantage in the fast-paced business environment, as they are constantly changing in response to the shifts within their industries and environments (ibid). In this section, the focus is to evaluate the main features of inquiring organizations, such as the focus on the acquisition of new knowledge, knowledge sharing, and the use of technologies like artificial intelligence (AI) and big data in the decision-making and application of knowledge.

The Continuous Learning Process in Inquiring Organizations

A key characteristic of inquiring organizations is that they are continually learning (ibid). Here, learning is not an event limited to courses or one-shot workshops but a social practice. Lifelong skill enhancement is encouraged by the inquiring organizations so that the employees update themselves concerning the industry standards, current trends, new technologies, and the best practices in the market. Education is viewed as an essential component of personal and organizational development. This environment also fosters the acquisition of soft skills including critical thinking, problem-solving skills, and flexibility (Grimsdottir & Edvardsson, 2018). People are free to think in new ways, generate hypotheses, and consult with others to enhance their collection of knowledge.

In many inquiring organizations, basic learning is offered through programs like training, workshops, and e-learning (Lin et al., 2015). Today, companies such as IBM and Google provide extensive learning management systems where people can find materials aimed at enhancing their skillsets. Furthermore, employees are normally encouraged to pursue their learning in areas of interest which makes them gain mastery in those areas (Nonaka, 2009). Another important component of continuous learning in inquiring organizations is the cultivation of a 'growth mindset' (Lin et al., 2015). Moreover, employees should not perceive certain difficulties as threats that hamper productivity, but as positive factors that foster development (ibid). This does not only increase motivation but also introduces a spirit of optimism, which ensures that employees learn how to cope with the difficulties and succeed even in the adverse environment

Cooperative organizations opt to fund knowledge acquisition activities where information can be obtained from sources including outside research, customer relations, and other departments (ibid). This allows the organization to be flexible and ready to accommodate new opportunities in the market or potential threats too. Inquiring organizations make it their responsibility to ensure that the people working for them are knowledgeable, capable, and ready to handle the problems that arise in a constantly changing environment (ibid).

Knowledge Sharing in Inquiring Organizations

Inquiring organizations appreciate that knowledge sharing is critical to creating value, increasing productivity, and addressing challenges (Lin et al., 2015). Such organizations foster conditions in which employees can freely share their ideas, concerns, or concerns and work together to address them. In this sense, the practice of knowledge sharing in inquiring organizations may be supported by a culture of openness (ibid). People are allowed and even encouraged to contribute their knowledge, ideas, and personal stories to the organization. It promoted trust in organizational processes, cooperation, and commitment to the success of the organization as a whole. For example, employees are offered tools such as Slack, Microsoft Teams, or internal message boards where they can freely engage in communication and exchange information (Grant, 2006). Besides informal communications, inquiring organizations tend to have more formal ways of knowledge storage that are databases, intranets, and wikis which contain the information, findings from researcher and the best practices of the organisations and can be accessed by users at different organizational levels (Lin et al., 2015). These repositories provide the organization with the institutional memory which enables it to harness and advance on past knowledge as opposed to 'reinventing the wheel'.

This implies that information sharing is not restricted to a specific department or a particular group of people. Investigative organizations encourage multidisciplinary integration to ensure relevant personnel from different departments (marketing, engineering, operations) work in one team. This management of diverse perspectives fosters out-of-the-box thinking and the creation of innovative solutions (Nonaka, 2009). Inquiring organizations also provide employees the freedom to create knowledge networks both within the organisation and externally. Such networks offer a broader range of specialists and perspectives, enhancing the flow of ideas and information.

Through knowledge sharing, inquiring organizations allow employees to tap into other perspectives of the organization and engage in problem-solving, which is more effective in generating improved decisions and cumulative creativity (Lin et al., 2015).

Inquiring Organization Technology Integration

The technology, especially AI and big data analytics, implements a crucial role in the knowledge processes of inquiring organizations (ibid). These technologies allow organizations to analyze data, detect patterns, and solve problems in a much quicker and smarter way. Big data is another area where AI tools are applied in the analysis of large data sets and providing insights for decision making. For instance, it is possible to use AI algorithms to study customer behaviors to detect fresh patterns, which can be crucial for organizations. The organizations that inquire use AI not only to manage the low-level tasks but also to reveal the knowledge that can remain unknown otherwise. For example, Amazon deploys AI to facilitate supply chain management, product recommendation, and targeted advertising and promotions (Brynjolfsson & McAfee, 2014).

Data mining entails the evaluation and analysis of large datasets from various sources such as social media, customers, and sensors. These insights prove beneficial for inquiring organizations to make sound decisions, enhance products, and streamline operations. For instance, healthcare industry firms apply big data analytics to enhance the patient experience and outcomes, cut costs, and forecast disease epidemics (Grimsdottir & Edvardsson, 2018).

The use of AI and big data in the knowledge environment of inquiring organizations makes the organizational decision-making process faster and more effective (Lin et al., 2015). For example, with the help of predictive analytics, it is possible to predict the demand for certain products to plan production and order goods accordingly. Real-time analytics also helps organizations to identify potential problems or issues in the process and solve them as they are noticed (ibid).

AI and automation tools benefit inquiring organizations by enhancing their efficiency and minimizing the necessity for human intercession (ibid). For instance, in the financial sector, AI is employed in transaction processing, fraud detection, and risk management. This allows the employees to handle other tasks that are more demanding hence increasing the productivity (Brynjolfsson & McAfee, 2014). Here, AI and big data help the inquiring organizations to make better decisions, increase their operational efficiency and strengthen their competitive positions. These technologies are not only useful for daily operations, but also for the process of learning and improvement due to fluctuating circumstances in the market.

USES OF SCIENTIFIC KNOWLEDGE

IBK (inquiry-based knowledge) is an important approach that makes organizations to ask questions, look for more knowledge and make changes based on what they have discovered (Davenport & Ronanki, 2018). This approach, based on the continuation of training and the implementation of new knowledge and innovative technologies (ibid), has extensive relevance in various sectors: healthcare, education, etc. Here, we discuss how inquiring organizations use IBK principles in the context of the healthcare and education sectors to improve decision-making, creativity, and organizational performance.

In the modern world, the healthcare industry has incorporated the use of inquiry-based knowledge in its desire to enhance patient care and organization effectiveness (Wickramasinghe et al., 2015). Through IBM Watson and other AI based systems have greatly impacted on the healthcare sector as it has empowered the healthcare practitioners to make informed decisions (ibid). These systems use big data to integrate patient data, clinical research data, and real-time data and offer recommendations to the clinical personnel. For instance, Watson for Oncology helps physicians in utilizing data to come up with a tailored care plan. Such systems analyze thousands of research articles, clinical trials, and patient outcomes to provide recommendations based on the scientific literature. This makes decision-making more objective, precise, and fast, which may benefit the patients (Davenport & Ronanki, 2018).

In the education sector, knowledge through inquiry is being used to help in shaping learning for students as individuals (Välimaa et al., 2016). The use of technology in the form of online curriculum and learning management systems (LMS) have become critical in facilitating differentiated instruction, fostering student interaction, and enhancing student achievement. With the help of advanced technologies such as Coursera, Khan Academy, and adaptive learning software, teachers are able to meet the needs of students with different learning styles. These platforms employ artificial intelligence that helps identify students' learning preferences, abilities, and difficulties, providing content that meets the needs of each learner. Such an approach aids in enhanced understandings, fosters independence among students, and enables instructors to meet the individual needs of the learners (Al-Qahtani & Aksoy, 2022). It is equally important to note that apart from delivering individualized instruction to students, technology integration in education offers insights to the instructors as well. With the help of analytics, teachers can observe the learners' achievements, realize the areas of their difficulties, and change the teaching approaches. This makes teaching and learning more responsive and flexible to meet the needs of the learner and the society.

Technology also supports collaborative learning by enabling students to complete assignments in groups, provide feedback to their peers, and share information in online classes (Välimaa et al., 2016). This collaborative environment enables acquisition of knowledge through sharing and different perspectives by the students. For example, Google Classroom and Microsoft Teams allow teachers and learners to share messages and collaborate on tasks and assessments. Such educational knowledge in terms of inquiry-based learning is similarly supported by the global learning communities within the digital platforms. The students in different parts of the world can interact through online forums, share the ideas, and work on the assignments. This exposure to different opinions helps students improve their learning processes and gain a more global perspective on certain problems.

Issues in Inquiry Organizations

A number of difficulties can be observed when implementing the principles of inquiry-based knowledge in organizations. Less willingness to change is one of the challenges that inhibit the integration of Networked Knowledge Environment (NKE) practices while the absence of technological infrastructure is also a challenge that hampers the integration of NKE practices (Lin et al, 2015). These challenges may constrain the optimal implementation of knowledge management strategies and tools hence impacting on the organizational performance and competitiveness.

Resistance to Change

Some of the major issues that organizations encounter when they intend to inquire include resistance to change (ibid). This is usually the result of many reasons such as organizational culture, employee attitude, and fear that is often instilled by a new idea or change. Workers are likely to resist change and will not embrace new technologies or methods if they feel that they do not require such changes in order to get the job done. It is often observed that in many organizations, especially in organizations where certain practices have been in the course for quite a long time, there may be certain cultures prevailing which go against change. Many receptionists coming from a more conventional background may not see new technologies or approaches as useful additions but as threats to their work. To overcome this resistance, it is crucial to have leadership commitment, communication, and showing the importance of change (Schein, 2018). Ignorance can also be another form of resistance where people may not understand new tools or methods. Lack of self-competence in using new technologies or new platforms may cause the employees to resist those changes. Hence, the inquiring organizations have to dedicate their efforts and resources in ensuring that the staffs is capable of operating these new systems. Another key area that is critical to address in order to overcome resistance is the understanding of change management. Managers have to not only promote the use of new technologies, but also engage employees in the process, explaining to them why such changes are necessary and how they will be beneficial for completing work (von Krogh et al., 2012).

Absence of Technological Support

Another important factor is the absence of relevant technology support (Schein, 2018). Some organizations may not have adequate capital to invest in the latest knowledge management technologies, including artificial intelligence, big data analysis, or shared platforms. As for the disadvantages, the costs of the technology that an organization needs to acquire may be high, and if an organization does not have enough money, it may not be able to afford the technology or, worse, sustain it. This can lead to gaps between organizations where the ones with more resources are able to implement NKE principles compared to those lacking such support structures. Thus, while organizations may have access to highly developed technology, the process of addressing these issues can be intricate and lengthy due to the need for integrating new systems with existing environments. It might also prove challenging to integrate legacy systems with new technologies since the former may not be compatible with the latter, and this would require additional investment in upgrading or replacing old systems (Schein, 2010).

KEY DRIVERS OF KNOWLEDGE MANAGEMENT (KM)

KM is a critical function that assists organizations in acquiring, organizing, and disseminating data as effectively as possible (Lin et al., 2015). KM enablers on the other hand are the conditions that facilitate the adoption and implementation of effective knowledge management practices by an organization. These drivers can be internal to the organization or external and affect both the structural and cultural aspects of knowledge sharing in business. Internal drivers are mainly associated with leadership and management processes, while external drivers involve technological and policy factors. In this section, we will look at the internal and external factors that drive KM, including leadership, knowledge repositories, technological development, and regulations.

Internal Factors That Influence Knowledge Management

Leadership

Some of the internal factors that organisations should consider for the success of KM practices include leadership (Pellegrini et al., 2020). Innovative and strategic-minded leaders who recognize knowledge as a critical organizational asset are key to cultivating a culture that supports knowledge acquisition and distribution. In a nutshell, per Schein (2010), organizational leaders have the primary responsibility of owning culture while simultaneously promoting the creation, sharing, and application of knowledge as the organization seeks to attain its vision and mission. Top management support: Successful KM initiatives start with a coherent and persuasive vision from senior management. It conveys the perception that knowledge is a valuable asset and informs the manner in which knowledge will be leveraged in the enterprise. The champions of knowledge management foster the purposeful coordination of stake holders and the employees to ensure that the whole company is integrated to work toward optimum utilization and advancement of the intellectual capital that is in the possession of the firm.

Moreover, leadership is a critical factor in the development of organizational culture that promotes transparency and cooperation (Pellegrini et al., 2020). Managers who lead by being open and inclusive also achieve the same in their organizations. The focus on knowledge management also increases the collaboration across the organisation and reduces the departmental silos; as a result, the employees can work in teams, share their knowledge, and come up with new solutions. It is fundamental for the execution of KM strategies to be underpinned by a culture of trust and cooperation. Another important component of leadership in knowledge management is the ability to ensure that learning is a continuous process. Strategic executives apply their resources in learning and skill building, training, and seminars to enhance their knowledge. When employees are motivated to acquire new knowledge and skills, organizations develop a KM environment where workforce is eager to contribute to overall organizational goals.

Knowledge Repositories

Another internal factor that fuels KM is knowledge repositories (Taskin & Van Bunnen, 2015). Such repositories consist of databases, document management systems, and knowledge-sharing platforms where crucial organizational knowledge resides and is disseminated to the workforce. It's important to note that as organizations continue to store large quantities of data, knowledge management provides the necessary framework for structuring and making this knowledge easily accessible when required (ibid). Characterized by Alavi and Leidner (2018), well-structured knowledge repositories facilitate the identification of the relevant information in timely manner aiming to improve the decision making processes and organization's performance.

Repositories of information tend to be centralized and more structured, helping the employees to find the needed documentation, information, and other materials (Taskin & Van Bunnen, 2015). There are centralized storage centers, including intranets and cloud-based systems, where employers can store information and allow employees to find it without browsing through several channels and systems.

This organization enhances the flow of work and diminishes the amount of time spent searching for materials. High fidelity knowledge management systems feature powerful search capabilities, metadata annotation and classification systems where employees can search for a particular subject, document or data easily (Taskin & Van Bunnen, 2015). By categorizing knowledge and using meta tags to add relevant keywords, organizations guarantee that workers, who are searching for the best solutions, will discover the most appropriate materials (ibid). Availability of these resources is a major factor that contributes to the success of KM. Collaborative knowledge-sharing platforms like SharePoint or Confluence can augment knowledge repositories. It is important to note that these platforms enable employees to share their knowledge in different forms such as documents, articles, and discussions, and offer an environment whereby users can engage and share their expertise and ideas. This way, the information is not only archived but constantly discussed and updated by the community members themselves.

Factors Outside Organizations that Influence Knowledge Management

Technological Advancements

Technological factors are perhaps the most significant external factors affecting KM since they provide the means of achieving real-time data acquisition and distribution (Wickramasinghe et al., 2015; Jarrahi et al, 2023). AI, IoT, and Cloud computing have brought dramatic change in the management of knowledge within organizations; they have enhanced the speed, efficiency, and access to knowledge (ibid). According to (Grimsdottir & Edvardsson, 2018), it is noteworthy that these technologies have allowed organizations to capitalize on the abundance of data and generate new, more fluid, and even more synchronous knowledge management systems. AI has also impacted the management of knowledge in organisations by introducing a way of collecting and analysing data and information in an automatic manner (Jarrahi et al, 2023). AI models can analyze large volumes of data, find patterns and produce insights that can be disseminated across the enterprise. This capability does not only enable the management of knowledge more efficiently, but also the identification of new sources of knowledge that may lead to innovation and better decisions. IoT is the integration of smart devices that enable real-time data acquisition and monitoring of events. Current examples span across the manufacturing, logistics, and healthcare fields where IoT devices gather information from sensors attached to machinery, equipment, or patients, respectively. These results are then disseminated throughout the organization, providing real-time information on operations, patient status, or equipment functionality. Another advantage of IoT data sharing is that it is done in real-time, and thus, it improves operational performance and enables fast decision-making – a consideration that is vital to organizations that want to stay relevant. Other advancements that have significantly impacted knowledge management include cloud computing whereby data can be accessed in real-time from anywhere. Cloud-based systems allow for the exchange of documents, projects, and knowledge between teams which are often situated in different locations. The cloud also means that information is updated in real-time, which means that employees are always informed, thus improving on communication and decreasing on time delays on decision making. Consequently, cloud-based technologies have emerged as an essential pillar of modern KM systems.

Regulatory Compliance

Another motivation for implementing KM is compliance with external regulations and relevant standards (Brynjolfsson & McAfee, 2019). Most industries are bound by laws and regulations that regulate how information ought to be managed, stored and disseminated. For instance, the GDPR in the EU and similar laws in other countries have elevated the levels of risk concerning data management in organizations. According to Brynjolfsson and McAfee (2019), one of the external pressures that lead to the development of strong KM systems is the pressure from regulators.

This is why regulations such as the General Data Protection Regulation (GDPR) make it mandatory to store and process customer data securely (ibid). Non-compliance with these regulations attracts severe penalties in terms of fines and reputational loss. Thus, organizations are forced to implement KM systems that monitor, archive, and process data while adhering to the regulation. They may contain elements like data encryption, secure access, and audit trails, which guarantee that the information is safe and can be tracked. In the healthcare and financial sectors, for example, there are industry-specific standards for the handling and dissemination of knowledge. For instance, in the United States, the Health Insurance Portability and Accountability Act (HIPAA) requires secure storage and transfer of patient information Organizations in these industries are therefore required to deploy KM systems that address these legal requirements regarding the security of the knowledge. In industries that have well-defined legal requirements, the KM system assists organizations in mitigating legal risks by ensuring that knowledge is properly captured. Contract management systems, IP management, and records of compliance documents can assist legal teams in avoiding the risk of litigation and maintaining legal compliance (ibid).

KM enablers, whether internal or external, serve a significant function in the management of knowledge within organizations (Allameh et al., 2011). Organizational initiatives such as leader support and information centers are important to foster knowledge sharing and guarantee that valuable know-what knowledge is available.

On the other hand, external factors like technology and legal demands force organisations to implement better and more efficient KM systems in order to stay relevant and adhere to laws. The synergy of these drivers forms an open KM ecosystem that provides opportunities for organizations to enhance and expand their KM initiatives and their benefits on innovation, operational effectiveness, and compliance.

HOW TO BECOME A KNOWLEDGE-BASED ENTERPRISE (KBE)

There are three significant steps to the transition to the KBE model (Lin et al., 2015; Alavi and Leidner; 2018; Davenport & Ronanki, 2018):

Establishing the Framework for Knowledge Management

The key component of any knowledge-based organization is the organizational infrastructure that supports the creation, storage, and sharing of the organizational knowledge. Strong IT supports allow for easy retrieval of valuable information and enhance the employee's ability to perform their tasks. The development of EM and KM one of the most important components of this infrastructure involves the establishment of knowledge repositories that are used to store information, documents, data and other knowledge resources. Alavi and Leidner (2018) posit that organizations that support knowledge repositories facilitate easy and fast access to information that enhances decision-making and operations.

Adopting effective IT systems like content management systems, enterprise resource plans, and knowledge management systems (like Microsoft SharePoint, Confluence, etc.) is a significant leap towards attaining KBE status. Such platforms enable knowledge to be stored and shared in a central and coordinated manner within an organization, irrespective of its level or department. These systems also offer components like document versioning, collaboration, and search functionality that allow employees to engage with and find knowledge assets.

Another important advancement in creating the foundations of a knowledge-based company is the transition to the cloud. Cloud computing provides up-to-date information and knowledge at workplace and hence, the employees may work from anywhere and at any time. This flexibility contributes to increased cooperation between teams that may be located in different regions as they can view, edit, or even work on the same documents or projects at the same time. Thirdly, cloud platforms offer flexibility, and this means that regardless of the growth of the organization, the knowledge management systems can also grow. Another essential aspect of knowledge management is the integration of multiple data sources and rationalization of data flows to guarantee that knowledge is timely, accurate, and easily accessible.

Fostering Collaboration

Another important aspect that can be identified in the discussion of the transition to a knowledge-based economy is the promotion of cooperation. Communicating and working together for the exchange of information is essential for enhancing creative processes and managing organizational knowledge effectively. Another outstanding approach is to use organizational structures that allow workers to share their knowledge and cooperate in achieving goals, such as group assignments and knowledge-sharing websites.

Leadership is the cornerstone of ensuring that everyone is on board in creating the culture of collaboration. Managers need to lead by example and use communication strategies that include information sharing, feedback, and promoting discussion. Furthermore, there is a need for organizations to have structures or procedures that support teamwork. This includes multidisciplinary working groups, task forces, project teams, and other circumstances that facilitate interaction between co-workers. First, it fosters a culture of cooperation, which helps in the sharing of information as well as problem-solving among the employees. Nonaka, 2019) further argue that organizational culture has a significant influence on knowledge creation as people are more willing to share their ideas when the environment is free of competition and judgment.

Technology platforms are critical in supporting collaboration. Apps like Microsoft Teams, Slack, and Google Workspace enable effective teamwork and real-time interactions irrespective of team members' locations. These platforms enable the exchange of ideas, sharing of documents, and even meetings among the employees. Moreover, most of these tools are compatible with knowledge databases thereby creating a central point to store and share the organization's knowledge. Promoting the use of these tools facilitates the dissemination of information, guarantees that employees get relevant information, and ensures that all the stakeholders are on the same page.

Another effective strategy for promoting collaboration is the formation of communities of practice (CoP). These are associations of employees with similar interest or with similar specialties or fields. CoPs make it possible for the employees to learn from one another, exchange ideas, and find solutions to some of the issues that arise in their workplace. In any case, whether they are aimed at particular technical areas, professional development, or organizational issues, CoPs foster knowledge growth and sharing. In this way, organizations not only help their employees develop their leadership skills but also tap into the expertise and knowledge of their community members.

Cross-functional integration enables the people in distinct departments or organizational divisions to combine their efforts to work on projects or undertake some problem-solving activities. One of the benefits of having a diverse workforce is that people from different backgrounds are able to brainstorm better and are likely to come up with unique ideas that would not be arrives at through idea isolation. Integrative communication also guarantees that ideas from all sectors of the establishment are shared and harmony is achieved, thus enhancing the chances of coming up with the best decisions.

Teamwork is one of the fundamental values of any knowledge-based organization. Hence, by developing an organizational culture, equipping them with the necessary tools and channels, and encouraging cross-functionality, it is possible to create an environment where new knowledge is widely disseminated and utilized in reaching the intended goals.

Measuring KM Effectiveness

To achieve this, organizations need to develop methods of measuring KM effectiveness in order to know whether the efforts being made are yielding the required results. Measurement helps in determining whether the KM strategies are being effectively deployed and how efficiently knowledge is being created, shared and utilized. Such measures assist organizations in determining their relative strengths and weaknesses, and they can modify their plans accordingly to meet organizational goals and objectives effectively.

Another important factor in evaluating the success of KM is the level of engagement by the employees. Employees who are engaged are more likely to share knowledge and participate in various activities within an organization or organization units. The degree of engagement can be assessed by surveys, feedback and involvement in the KM activities by the employees and the overall increase in the frequency of participation in knowledge management activities. Greater utilization means that the employees are realizing that the knowledge management systems are beneficial and are actively participating in the development and sharing of knowledge.

Another measure is that of knowledge reusability where one can determine how often knowledge is applied. This metric assesses how often knowledge that has been created or stored in the organization's knowledge repositories is accessed, modified, or utilised by the employees. A high level of reuse means that knowledge management systems facilitate the dissemination of knowledge that can be utilised by individuals in their day-to-day tasks. Efficiency is also improved because employees do not have to start researching from scratch or even go over what others have already done. Measuring knowledge reuse is another aspect that can be evaluated by assessing the number of document downloads, shared resources and how often the knowledge management tools are used.

KM initiatives are developmental in nature with the aim of promoting innovative ideas and solutions through pooling of employees' resources. Hence, it is recommended that organizations measure innovation-related parameters like the volume of new ideas, patents registered, or products created by leveraging the knowledge deficit. The effectiveness of KM in implementing innovation can also be evaluated in terms of the rate at which new solutions are rolled out and how effectively they meet customer requirements. Last but not the least, KM strategies should be measurable and improve the overall performance of the organization. Measures like customer satisfaction index, the time taken in introducing new products or services, and overall organizational performance may be used to measure the effectiveness of KM interventions. The key benefits of KM are the improvement of product quality, customer service, and organizational efficiency, all of which contribute to organizational effectiveness.

It is also important to evaluate the flow of knowledge within the organization in order to determine the effectiveness of KM. Companies should also monitor the flow of knowledge across departments, teams, and employees and the rate and reliability in sharing such information. Proper knowledge flow indicates that the KM system is performing optimally and that employees are participating in the knowledge network.

To successfully transform into a Knowledge-Based Enterprise, organizations need to invest in appropriate IT structures and cultures and develop methods of evaluating the effectiveness of knowledge management. Establishing a strong and efficient IT foundation guarantees the availability and accessibility of knowledge and unleashing the collaboration via tools and culture strengthens the innovation and problem-solving skills. Last of all, implementing the measures of KM effectiveness, including employee engagement, rates of knowledge reuse, and innovation assists organizations in evaluating their performance and make necessary adjustments. Incorporating knowledge management into the DNA of an enterprise fosters growth and increases efficiency while sustaining the competitiveness of the enterprise given that the world is gradually shifting into a knowledge-based economy.

CONCLUSION

Knowledge, in an economic sense is a unique construct as it has increasing rather than diminishing returns. This means that as knowledge is shared it does not reduce over time but increases in value. This unique aspect of the knowledge construct makes the need to harness knowledge from data and information a key strategic necessity for all healthcare organisations today.

In this chapter, the reader has been introduced to the concept of the NKE and the significance of this framework in the modern business world. With organizations struggling to cope with the changes in the global environment, it has become extremely crucial to embrace knowledge based strategies. Therefore, through the use of technological resources, organizational learning, and various forms of collaboration, NKE provides strategies to create sustainable competitive advantages.

In order to set the context for the discussion, the chapter started with the identification of the main features of networked knowledge environments: collaboration, decentralization of information, and real time access to knowledge. Such applications as Microsoft Teams, Slack, and cloud solutions allow the employees to communicate with each other, share knowledge, and work together without leaving their workplace. Due to the decentralization of information and the integration of real-time data exchange, organizations can become more flexible and adapt quickly to the new conditions of the market. These aspects of NKE make businesses not only more efficient, but also more prepared to harness knowledge as a valuable resource.

Additionally, the chapter expanded on such an idea as the inquiring organizations that focus on learning and knowledge acquisition. These organizations promote a knowledge-sharing culture where employees are free to share and exchange knowledge, solve problems, and generate new knowledge. The SECI model by Nonaka (2009) was mentioned as a framework that facilitates conversion and utilisation of knowledge for the purpose of organisation development. This highlighted the need for both human and technological components, and positioned AI and big data analytics as essential tools for improving decision-making and organizational outcomes.

The examination of knowledge management (KM) drivers was also conducted in this chapter. The key internal enablers include visionary leadership as well as knowledge repositories, which are crucial in developing comprehensive knowledge management practices. Technological enablers like AI & IoT and regulatory factors like GDPR are external forces that are influencing the management of knowledge within organizations today. To elaborate on these drivers, the following was a breakdown of how these concepts are applied in the business world using the cases of Amazon and Microsoft. By employing advanced technology solutions and promoting the exchange of knowledge, businesses can enhance their overall performance and cultivate a more effective and creative labor force.

The chapter then proceeded to provide a clear roadmap for the establishment of the knowledge-based enterprise. Among the initial steps, creating a solid foundational framework, including knowledge repositories and cloud resources, was listed. Such changes may include encouraging team-based projects, promoting the use of knowledge-sharing platforms, and more. This also introduced the significance of evaluating the effectiveness of KM, where some of the parameters that may be used include the level of engagement of employees, rate of reuse of knowledge, and the outcome of innovations.

In conclusion, the networked knowledge environment is imperative for organizations in the current business environment because it helps organizations to be more competitive. Thus, organisations that incorporate IT, teamwork, and best practices in knowledge management are poised to thrive in the emerging knowledge economy. In this way, they are able to improve their internal capacity while at the same time providing improved and more competitive products, services and customer experiences in a global context. What is important to understand here is that knowledge-driven change is not a choice but a necessity for those who want to continue being players of a game.

REFERENCES

Al-Qahtani, M. A., & Aksoy, M. S. (2022). Collaborative tools and techniques of knowledge sharing: A literature review. *Journal of Computer and Communications, 10*(5), 45–54. DOI: 10.4236/jcc.2022.105003

Alavi, M., & Leidner, D. E. (2018). Knowledge management systems: Conceptual foundations and research issues. *Management Information Systems Quarterly, 42*(1), 107–136. DOI: 10.2307/3250961

Allameh, S. M., Zare, S. M., & davoodi, S. (2011). Examining the impact of KM enablers on knowledge management processes. *Procedia Computer Science, 3*, 1211–1223. DOI: 10.1016/j.procs.2010.12.196

Brynjolfsson, E., & McAfee, A. (2014). The second machine age: Work, progress, and prosperity in a time of brilliant technologies. WW Norton & company.

Davenport, T. H., & Prusak, L. (1998). *Working knowledge: How organizations manage what they know*. Harvard Business School Press.

Davenport, T. H., & Ronanki, R. (2018). Artificial intelligence for the real world. *Harvard Business Review, 96*(1), 108–116.

Grant, R. M. (2006). The knowledge-based view of the firm. *Journal of Management Studies, 55*(7), 1050–1073.

Grimsdottir, E., & Edvardsson, I. R. (2018). Knowledge management, knowledge creation, and open innovation in Icelandic SMEs. *SAGE Open, 8*(4), 2158244018807320. DOI: 10.1177/2158244018807320

Hoadley, C., & Kali, Y. (2019). Five Waves of Conceptualizing Knowledge and Learning for Our Future in a Networked Society. In Kali, Y., Baram-Tsabari, A., & Schejter, A. M. (Eds.), *Learning In a Networked Society. Computer-Supported Collaborative Learning Series* (Vol. 17). Springer., DOI: 10.1007/978-3-030-14610-8_1

Jarrahi, M. H., Askay, D., Eshraghi, A., & Smith, P. (2023). Artificial intelligence and knowledge management: A partnership between human and AI. *Business Horizons, 66*(1), 87–99. DOI: 10.1016/j.bushor.2022.03.002

Lin, C. H., Wei, A., Yang, H., Pittayachawan, S., Vogel, D., & Wickramasinghe, N. 2015, January. Inquiring knowledge management systems- a Chinese medicine perspective. In *System Sciences (HICSS), 2015 48th Hawaii International Conference on* (pp. 3682-3690). IEEE.

Loureiro Cardoso, T. M., Pinto, J. P., & Pestana, F. (2024). Networked research and open science: The WEIWER® experience. *Educational Media International, 61*(1–2), 16–25. DOI: 10.1080/09523987.2024.2357475

Nonaka, I. (2009). *The knowledge-creating company revisited.* Oxford University Press.

Pellegrini, M. M., Ciampi, F., Marzi, G., & Orlando, B. (2020). The relationship between knowledge management and leadership: Mapping the field and providing future research avenues. *Journal of Knowledge Management, 24*(6), 1445–1492. DOI: 10.1108/JKM-01-2020-0034

Schein, E. H. (2010). *Organizational culture and leadership* (5th ed.). Wiley.

Taskin, L., & Van Bunnen, G. (2015). Knowledge management through the development of knowledge repositories: Towards work degradation. *New Technology, Work and Employment, 30*(2), 158–172. DOI: 10.1111/ntwe.12049

Välimaa, J., Papatsiba, V., & Hoffman, D. M. (2016). Higher Education in Networked Knowledge Societies. In Hoffman, D., & Välimaa, J. (Eds.), *RE-BECOMING UNIVERSITIES? The Changing Academy – The Changing Academic Profession in International Comparative Perspective* (Vol. 15). Springer., DOI: 10.1007/978-94-017-7369-0_2

von Krogh, G., Nonaka, I., & Rechsteiner, L. (2012). Leadership in organizational knowledge creation. *Journal of Management Studies, 55*(1), 52–78.

Wickramasinghe, N., Haddad, P., Han-Lin, C., Moghimi, H., Delimitros, H. And Vaughn, S. 2015. *Identifying key success factors for the adoption and implementation of CPOE Systems into the Australian private healthcare sector.* Presented at AMCIS 2015.

Chapter 2
The Construct of Knowledge

ABSTRACT

The knowledge construct has unique economic properties. Namely, it is one of very few constructs that possess the quality of increasing not diminishing returns. This means that knowledge can be shared and added to and never decreases in value. Further, the knowledge construct is complex and has tacit and explicit properties, is found in meaning and created in sense making activities. In healthcare knowledge can reside in patients as well as in experts such as clinicians. Naturally, the types of knowledge are different but the exchange of these different types of knowledge is essential to superior superior clinical decision making. The following serves to demystify the knowledge construct.

INTRODUCTION

Knowledge is a fundamental component of the modern information society wherein it plays a critical role in supporting decision making, introducing innovations, and developing strategies at all levels (Wickramasinghe, 2006). Appropriate knowledge management strategies are pivotal to enabling these processes and to ensure that the identified knowledge is properly implemented to improve the performance of an organization (Zipperer, 2016). In the context of healthcare, professionals use a broad spectrum of knowledge base, ranging from evidence-based protocols and scientific discoveries to practical expertise and patients' preferences to enhance the sharing of information between different stakeholders and consequently improve patient care and health system efficiency (Kothari et al., 2011; von Lubitz and Wickramasinghe, 2006). This chapter will expand on this aspect of knowledge and present key concepts such as definitions and types of knowledge as well as the key stages of the knowledge management cycle in the healthcare context.

DOI: 10.4018/979-8-3373-1877-6.ch002

UNDERSTANDING KNOWLEDGE

Knowledge is a complex phenomenon, which has been defined and conceptualized in many different ways in different fields and traditions (Wickramasinghe, 2006). Knowledge can be defined as the understanding, appreciation, and practical experience that people and groups gain from learning, practice, and exposure to information.

Epistemologically, the definition of knowledge has been expressed as "justified true belief," (JTB) (Parikh & Renero, 2017) which incorporates belief, truth, and justification. The JTB account of knowledge is a traditional view that posits that for one to have knowledge, three elements have to be met: The proposition that one has the knowledge has to be true; the individual possessing the knowledge has to believe the proposition; and the individual possessing the knowledge has to have justification for believing the proposition (ibid). However, there are problems with this definition perhaps best illustrated by Edmund Gettier who provided cases of justified true beliefs that could not be considered knowledge, hence the JTB definition is not enough (Ichikawa & Steup, 2018).

In the organizational context, knowledge is commonly identified as a competitive resource that allows organizations to make the right decisions, solve challenges, and produce more value (von Lubitz and Wickramasinghe, 2006; Wickramasinghe, 2006). It is the use of information, knowledge and experience in the context of decision making that makes knowledge a powerful strategic asset (ibid). It is also important to differentiate between different forms of knowledge, including explicit knowledge that is easily written down or explained, and tacit knowledge that is embedded in individuals and situations. Coordinating both types is mandatory to ensure an organization's efficiency and effectiveness (Nonaka & Takeuchi, 1995).

It is crucial to differentiate between the concepts of knowledge and other similar concepts such as data and information. Data is the unprocessed facts or pieces of information while information is the processed and presented data (Wickramasinghe, 2006). Knowledge, on the other hand, is the ability to use and blend information; it includes manual information, which is documented and shared, and implicit information, which is internal and intuitive (ibid). This hierarchy points to a growing significance and sophistication as one progresses from data to information to knowledge (Alavi & Leidner, 2001).

The concept of knowledge is multifaceted and made up of different kinds of knowledge (ibid). For instance, procedural knowledge means knowing how to do something and declarative knowledge means knowing that something is true (Wickramasinghe, 2006). Further, two well-known concepts in epistemology include a priori knowledge, which is knowledge that does not require prior experience, and a posteriori knowledge, which requires prior experience (ibid). These classifications aid in understanding the various forms of knowledge and how they can be utilized (Stanley & Williamson, 2001).

The process of knowledge management in organizations usually takes the following stages: generation, documentation, dissemination, and implementation (Alavi & Leidner, 2001). The role of knowledge management practices in enabling these processes is therefore critical in capturing and applying such knowledge to improve performance (ibid). In the context of healthcare, it is especially crucial to manage knowledge strategically. In this way, healthcare professionals use a broad spectrum of knowledge base, ranging from evidence-based protocols and scientific discoveries to practical expertise and patients' preferences (Wickramasinghe, N. & Davison, G, 2004). Knowledge management in health care has the potential for making patients' outcomes better, organizational processes more effective, and providing a stimulus for innovation in treatment and patient care (ibid). When healthcare organizations are able to integrate both explicit and tacit knowledge, they can be able to address the challenges that characterize modern practice while providing the best quality services (Wang et al., 2018).

TYPES OF KNOWLEDGE

Nonaka was one of the first scholars to clearly articulate the different types of knowledge (Nonaka & Takeuchi, 1995; Wickramasinghe, 2006; Nonaka & Toyama, 2003):

Explicit Knowledge

This is information that is written down, more easily recorded, and less likely to be held in people's heads. It is often stored in manuals, databases, and procedures within organizations. This type of knowledge is more transferable and it can be easily communicated and transferred from one person to another or from one team to another. For instance, a standard employee manual in a company or a manual on how certain machinery should be operated are examples of tacit knowledge.

Companies should possess explicit knowledge because of standardization and scalability in organizations. It makes knowledge and best practices easily available in the organization and reduces reliance on key personnel. Nonaka and Takeuchi (1995) defined explicit knowledge as knowledge that is created from the conversion of tacit knowledge through the process of externalization of ideas into documents. But the problem is to keep such knowledge current and thus usable for practical purposes.

Modern research focuses on the use of technology as a tool to manage explicit knowledge. For example, Alavi and Leidner (2014) find that knowledge management systems (KMS) are critical for the effective sharing of explicit knowledge. They can help organizations develop systems that allow easy storage of information in a central location and easy access to it.

Tacit Knowledge

Tacit knowledge is individual, situation-related and cannot be easily reduced to writing. This refers to an understanding and knowledge that people gain from practice in performing certain tasks. In contrast to the explicit knowledge, tacit knowledge is inextricably embedded within the individual and is difficult to codify. For instance, the skills of a particular manager to solve conflicts or a chef to prepare new recipes are examples of tacit knowledge.

Implicit knowledge plays an important role in innovation and problem solving as it is more intuitive in nature. According to Nonaka and Takeuchi (1995), the bulk of organizational knowledge is tacit in nature. But it can only occur through interpersonal communication, and, most importantly, through trust and common experiences. In the workplace, the best forms of knowledge transfer include; mentorship, apprenticeships, or working together as a team.

Past studies have also looked at the use of social networks in transferring tacit knowledge. For example, Inkpen and Tsang (2016) have also underscored the role of social capital in facilitating the transfer of knowledge within and between organizations. They stated that close interpersonal contact and trust are necessary for this sort of knowledge sharing.

Embedded Knowledge

This type of knowledge is well entrenched in organizational procedures, products and services, as well as culture and practice. It is the aggregate of experiences that have been institutionalized in the organization as knowledge assets. For example, the design of a product, how a particular process is conducted or the environment in which an organization operates are all examples of tacit knowledge.

This form of knowledge is fundamental but is often overlooked by organizations even though it has a causal link with the behavior and performance of these organizations. This means that work is not disrupted since the knowledge is not attached to specific people. On the same note, it has the potential of hindering change if the organizational culture is inflexible on some of the practices it encourages.

Procedural Knowledge

Procedural knowledge refers to the knowledge of how something is done or how to do things (Carpenter, 1986; Moses, 1994; Star, 2005). It develops slowly and unconsciously and often as part of the process of imitation. For instance, knowing how to operate a machine, how to manage the project or how to negotiate a contract is considered as procedural knowledge. This is important in doing work and realizing goals and objectives within the required time frame. It enables people to apply their skills and experience in a bid to achieve set goals and objectives. This is because as stated earlier procedural knowledge is less explicit than declarative knowledge because it concerns how things are done and how knowledge is obtained.

The distinctiveness between the categories of knowledge including explicit, tacit, embedded, declarative, and procedural knowledge should be clearly understood. Every type of knowledge plays its role in organizational behaviour, decision making and performance in a unique manner. Thus the knowledge mentioned above, organizations are able to enhance their competencies, generate new knowledge, and maintain competitive advantage in the current business environment.

ORGANIZATIONAL KNOWLEDGE LIFE CYCLE (OKLC)

Knowledge management is a corporate activity that involves identifying, acquiring, storing and using knowledge assets (von Lubitz and Wickramasinghe, 2006). This cycle is also referred to as the Organizational Knowledge Life Cycle and is a conceptual model that outlines a process through which an organization can leverage its knowledge assets and ensure that they become a source of competitive advantage and sustainability. As a result, the lifecycle could be divided into the following stages: Knowledge Generation and Acquisition, Knowledge Storage and Retrieval, Knowledge Transfer and Implementation, Knowledge Management and Maintenance, and Knowledge Assessment (Wickramasinghe, 2006). Each phase plays a crucial role in ensuring that knowledge is appropriately captured and utilized in achieving corporate goals. In the subsequent sections of this chapter, each stage will be explained in more detail and supported with examples from the contemporary sources to ensure the adequate level of familiarity with the topic (Wickramasinghe, 2006; Nonaka & Toyama, 2003).

Knowledge Creation

Knowledge Creation is the first step in the knowledge management process commonly referred to as the knowledge cycle (Ly et al. 2020; Popadiuk & Choo,

2006). It refers to creating new ideas, information, ideas, knowledge and experience through research, innovation and learning. Organizations promote knowledge development through the creation of environment that supports asking questions, testing ideas, and learning from other people. Nonaka and Takeuchi's (1995) SECI model (Socialization, Externalization, Combination, and Internalization) continues to be useful in explaining the processes through which tacit and explicit knowledge combine to generate new knowledge. However, the modern approaches stress on the use of technology platforms, open innovation, and multi-disciplinary teams to create new knowledge (Davenport and Prusak, 2013).

For instance, companies such as Google and 3M allow their employees to work on different ideas and projects for some hours in a week. This concept referred to as "20% time" at Google has led to innovations like Gmail, Google Maps (Bock, 2015). Moreover, knowledge creation is fostered by external relationships, including university and research institutes, and industry associations, which help in developing different ideas and viewpoints (Chesbrough, 2011).

Knowledge Capture

Once the knowledge is generated, it must be stored or recorded to avoid loss of such knowledge in the future. It entails the process of converting implicit knowledge (knowledge in the head) into explicit knowledge (knowledge on paper). Some of the methods used include documentation, storytelling, and the use of knowledge repositories.

For instance, organizations apply wikis, databases, and content management systems as tools for knowledge storage and organization. As for the fourth approach, storytelling is an effective means for capturing tacit knowledge since it enables employees to explain experiences and ideas using real-life examples (Swap et al., 2010). Also, with the help of artificial intelligence and machine learning, knowledge can be extracted from unstructured content such as emails, transcripts of meetings, and social media feeds (Hislop, Bosua, & Helms, 2018).

Knowledge Sharing

Knowledge Sharing is defined as the act of redistributing the knowledge that has been acquired within an organization (Wickramasinghe, 2006; von Lubitz and Wickramasinghe, 2006). Knowledge sharing helps disseminate knowledge to those who need it and it can be a means of promoting teamwork and innovation. This is achieved through training, workshops, collaboration arenas, and communities of practice.

The advancements in the application of communication and collaboration technologies, such as Slack, Teams, and SharePoint, have significantly transformed the facilitation of knowledge sharing across organizations or within departments and teams through real-time communication irrespective of the location of the employees (Leonardi, Neeley, & Gerber, 2012). Members of a community of practice (CoP), people with related interests or in the same line of business, are central to knowledge sharing. For instance, at NASA, knowledge management through CoPs has been used in the sharing of best practices and lessons learned between projects to avoid isolation of crucial information (Wenger, McDermott, & Snyder, 2011).

Knowledge Application

At the core of managing knowledge is the intention to apply it (Davenport& Prusak, 2013). This chapter asserts that shared knowledge is important to organizations as it enhances processes, informs decision-making, and fosters creativity. Knowledge application is usually in the form of practices, policies, procedures, and innovative approaches to issues (ibid).

For instance, Toyota's "Just-in-Time" production system is one of the best examples of knowledge application. Using the available knowledge on supply chain management and production, Toyota managed to minimize wastage and increase efficiency (Liker, 2011). Similarly, business intelligence and data analytics help organizations to incorporate knowledge into their decision-making processes to adapt to changing markets and consumers promptly (Davenport & Harris, 2017).

Knowledge Retention

Knowledge Retention is very important since knowledge is a corporate asset that needs to be protected and passed forward to other employees (Davenport& Prusak, 2013). This stage requires the updating of active repositories and the promotion of mentorship, as well as fostering an environment that promotes knowledge retention.

Of all the knowledge retention mechanisms, mentorship is most successful in helping organizations retain tacit knowledge because it involves the transfer of knowledge from the senior to the junior employees (ibid). Furthermore, knowledge management systems (KMS) are employed by organizations to capture and index knowledge so that it does not fall into the abyss when a worker is let go of. For instance, consulting firms such as McKinsey & Company, have created rich knowledge databases which contain lessons from previous projects, and which the consultants can use in future assignments (Hansen, Nohria, & Tierney, 2015).

Knowledge Evaluation

The last stage of the knowledge lifecycle is the Knowledge Evaluation (Davenport& Prusak, 2013), which entails determining the currency and reliability of the knowledge assets. It is essential for organizations to carry out the knowledge review continuously in order to update the knowledge base and to check that it is in congruence with the strategic direction of the organization.

WHY IS KNOWLEDGE IMPORTANT IN THE DELIVERY OF HEALTHCARE SERVICES?

Knowledge management is crucial in the healthcare sector because the understanding of disease and its treatment is instrumental in the management of patients (Von Lubitz & Wickramasinghe, 2006a). Healthcare as an ever-evolving and sophisticated field requires individuals and institutions to learn, organize, and use knowledge to enhance care. The value of knowledge in healthcare can now be discussed from several perspectives (ibid): Decision making, Safety and Quality, Organization and Management, Learning, Teamwork and Participation, and Patients. Each of these dimensions underscores the importance of knowledge in the reform of healthcare delivery systems and enhanced quality care. In the following sections, each of these dimensions will be further discussed Von Lubitz & Wickramasinghe, 2006a; Wickramasinghe, 2006a; Wickramasinghe & Schaffer, 2010).

Improved Decision-Making

It is crucial that healthcare workers keep abreast of the latest information in the medical field to be able to diagnose the patients, identify the right course of action, and effectively handle the patients' cases. For instance, the evidence-based practice (EBP) where decision-making in healthcare is made on the basis of clinical experience and the evidence available is a good example of how knowledge enhances decision-making (Sackett et al., 2010). EBP ensures that the clinical decisions being taken are informed by the best evidence and not by past practices or recommendations.

For instance, clinical decision support systems (CDSS) are IT systems that incorporate knowledge databases to provide advisories to clinicians on time. These systems deal with patient data and may offer advice on medication and other tests based on studies (Bright et al., 2012). Kawamoto et al. (2014) have stated that the integration of CDSS into EHRs improves clinical decision making and patient outcomes. Therefore, knowledge management systems play a central role in enhancing the quality of care by equipping the healthcare practitioners with the relevant information in their area of practice.

Enhanced Patient Safety

Patient safety is crucial in the delivery of care as it helps prevent errors and adverse events. Knowledge of how it works and its compatibility helps the health care practitioners to avoid adverse effects on their clients. For instance, the CPOE systems have the drug information and interaction/contraindication knowledge to alert the clinicians on the potential risks when prescribing drugs (Bates et al., 2014).

Furthermore, with KM systems, it is possible to disseminate information on safety procedures and policies. For instance, the World Health Organization's (WHO) Surgical Safety Checklist is a knowledge-based tool that has been demonstrated to reduce the incidence of surgical complications and post-surgical mortality because it ensures that there are steps that need to be followed before, during, and after surgery are followed (Haynes et al., 2011). When knowledge is incorporated into processes and processes into tools, then health care institutions can guarantee that errors are minimized.

Operational Efficiency

The improvement of operational efficiency is another area where knowledge management has a lot of benefits. Many healthcare organizations encounter problems in managing resources, patients' throughput, and appointments. It also allows these organizations to understand the best practices and standard operating procedures to enhance efficiency, eliminate inefficiencies, and offer quality services.

For instance, lean management systems that are grounded on understanding of process improvement has been integrated and implemented in the healthcare sector to address sources of waste (Graban, 2016). Healthcare organizations can use data analysis and bottlenecks to implement specific strategies for speeding up that workflow. Likewise, information regarding staffing density and patient circulation can be beneficial for eliminating waiting time gaps and enhance the patient satisfaction level (McHugh et al., 2011).

Additionally, KM can also help to enhance operational effectiveness by offering real time data and business intelligence. For example, predictive analytics tools can predict patient admissions based on the analysis of past data and help the hospitals to manage their resources better (Davenport & Harris, 2017). As a result, through the application of knowledge in an organization, healthcare delivery can be made more efficient and effective.

Continuous Learning and Improvement

It is essential to stay updated with the most recent advancements in the healthcare field since new research, technologies, and treatments are being developed at a faster pace. Effective knowledge management means that the professionals working in the healthcare sector are in a position to learn and update themselves, which will go a long way in improving the quality of the services offered to the patients. This is particularly important in an age where the approach to patient management is shifting more towards the precision medicine, in which the treatment is delivered according to the patient's genetic profile, environment, and lifestyle (Collins & Varmus, 2015).

This is how knowledge is transferred to practicing healthcare workers; through Continuing Medical Education (CME) programs and online learning platforms. These programmes update the clinician with current research findings, evidence based practice and current practices in their respective fields of specialty (Mazmanian, Davis, & Galbraith, 2011). Furthermore, knowledge management systems can monitor the efficiency of the training processes and determine what further lessons to provide to guarantee that healthcare workers can offer optimal care.

Facilitating Collaboration

Interdisciplinary cooperation is crucial in healthcare since patients require a number of different services from different specialists. Knowledge sharing creates a sense of teamwork by allowing individuals to exchange information, co-ordinate client care as well as design a comprehensive approach to the treatment. For instance, electronic health records (EHRs) contain the information of a patient and can be accessed and updated by different healthcare disciplines in the management of patient's care plan (Blumenthal & Tavenner, 2010).

Patient Empowerment

Educating patients about their disorders and medications enables them to take an active role in their medical processes and improve their health. Knowledge management in healthcare is incomplete without patient education as it empowers people to make informed decisions and follow prescribed treatment regimens. For instance, patient portals and mHealth applications offer a way through which patients can view their health records and interact with their caregivers (Kahn et al., 2014).

Another approach that involves the use of patient knowledge is shared decision-making (SDM). SDM is a process in which both the healthcare providers and patients engage in making decisions that are evidence-based and aligned with the patient's preferences (Elwyn et al., 2012). This has been found to enhance patient satisfaction and compliance with treatment, especially with long-term illnesses (Stacey et al., 2017). It will also help the healthcare organizations to boost their patients' engagement and therefore improve their outcomes.

It is crucial for healthcare organizations to integrate effective knowledge management practices in order to enhance the quality, safety, and efficiency of care. Knowledge management in healthcare improves decision making, patient safety, operational efficiency, continuous learning, collaboration, and patient engagement, which all lead to the improvement of the healthcare delivery system. However, knowledge is a strategic asset in the current dynamic and competitive healthcare environment.

CONCLUSION

It is important for organizations to understand the different types of knowledge so as to harness their intellectual capital efficiently. There are four main types of knowledge, namely, referential, experiential, procedural and contextualized knowledge which is all important in organizational settings. Explicit knowledge is more structured and documented thus can easily be passed from one person to another. For instance, manuals, procedures, and databases. This type of knowledge can be very easily disseminated and is usually documented in organizational memory. For example, SOPs, which stands for Standard Operating Procedures, are an example of explicit knowledge; these are detailed guidelines that employees have to follow whenever they are performing a certain task.

Tacit knowledge is individualistic, contextual and difficult to codify. It entails ideas, notions and feelings that people acquire through practice. The sharing of such knowledge is normally done face-to-face and this is because it is derived from one's own experience. For instance, practical experience, such as a master artisan's ability to make an intricate sculpture, cannot be learnt from a manual, but is learnt over time.

Tacit knowledge is also known as the "know-how," as it is the application of explicit knowledge. It occurs in a particular context, often unconsciously to the learner while he or she is being trained. This type of knowledge is often hard to capture but can be easily shared within an organization. For instance, a worker may discover how a sales pitch is developed from a manual and then develop new sales pitches based on that knowledge for different clients.

Embedded knowledge lies in the processes, products, culture, and activities that define the organization. It is the knowledge that has been institutionalized and embedded into systems and structures the way work is done within the organization. It is an implicit sort of knowledge that can be seen within how an organization and its members go about their work. For instance, the manner in which a firm organizes its products, or how it approaches quality assurance exhibits information-embedded knowledge.

The overall process of knowledge management, from its generation through to its use, is crucial for organizations that seek to create competitive advantage derived from their intangible assets. In healthcare, the role of knowledge is especially significant as it affects decision making, patient outcomes, productivity and change. This paper posits that through implementing comprehensive knowledge management, the service delivery of healthcare organizations and the general well-being of patients can be improved. For example, knowledge management can enable better clinical decision making since knowledge workers will be able to get updated information when they need it. Besides, this can enhance the integration of health care teams, leading to an improved quality of care for the patients.

In conclusion, realization on the various categories of knowledge and the application of appropriate knowledge management strategies are vital for the success of organizations especially the healthcare organizations to enhance performance and accomplish the desired results.

REFERENCES

Alavi, M., & Leidner, D. E. (2001). Review: Knowledge Management and Knowledge Management Systems: Conceptual Foundations and Research Issues. *Management Information Systems Quarterly, 25*(1), 107–136. DOI: 10.2307/3250961

Bates, D. W., Saria, S., Ohno-Machado, L., Shah, A., & Escobar, G. (2014). Big data in health care: Using analytics to identify and manage high-risk and high-cost patients. *Health Affairs (Project Hope), 33*(7), 1123–1131. DOI: 10.1377/hlthaff.2014.0041

Blumenthal, D., & Tavenner, M. (2010). The "meaningful use" regulation for electronic health records. *The New England Journal of Medicine, 363*(6), 501–504. DOI: 10.1056/NEJMp1006114

Bock, L. (2015). *Work Rules!: Insights from Inside Google That Will Transform How You Live and Lead.* Twelve.

Bright, T. J., Wong, A., Dhurjati, R., Bristow, E., Bastian, L., Coeytaux, R. R., & Lobach, D. (2012). Effect of clinical decision-support systems: A systematic review. *Annals of Internal Medicine, 157*(1), 29–43. DOI: 10.7326/0003-4819-157-1-201207030-00450

Carpenter, T. P. (1986). Conceptual knowledge as a foundation for procedural knowledge. In Hiebert, J. (Ed.), *Conceptual and procedural knowledge: The case of mathematics* (pp. 113–132). Lawrence Erlbaum Associates, Inc.

Chesbrough, H. (2011). Open services innovation: Rethinking your business to grow and compete in a new era. *Journal of Business Research, 64*(6), 689–690.

Collins, F. S., & Varmus, H. (2015). A new initiative on precision medicine. *The New England Journal of Medicine, 372*(9), 793–795. DOI: 10.1056/NEJMp1500523

Davenport, T. H., & Harris, J. G. (2017). *Competing on analytics: The new science of winning*. Harvard Business Review Press.

Davenport, T. H., & Prusak, L. (2013). *Working knowledge: How organizations manage what they know*. Harvard Business Review Press.

Elwyn, G., Frosch, D., Thomson, R., Joseph-Williams, N., Lloyd, A., Kinnersley, P., & Barry, M. (2012). Shared decision making: A model for clinical practice. *Journal of General Internal Medicine, 27*(10), 1361–1367. DOI: 10.1007/s11606-012-2077-6

Graban, M. (2016). *Lean hospitals: Improving quality, patient safety, and employee engagement*. CRC Press.

Hansen, M. T., Nohria, N., & Tierney, T. (2015). What's your strategy for managing knowledge? *Harvard Business Review*, *77*(2), 106–116.

Haynes, A. B., Weiser, T. G., Berry, W. R., Lipsitz, S. R., Breizat, A. H. S., Dellinger, E. P., & Gawande, A. A. (2011). Changes in safety attitude and relationship to decreased postoperative morbidity and mortality following implementation of a checklist-based surgical safety intervention. *BMJ Quality & Safety*, *20*(1), 102–107. DOI: 10.1136/bmjqs.2009.040022

Hislop, D., Bosua, R., & Helms, R. (2018). *Knowledge management in organizations: A critical introduction*. Oxford University Press. DOI: 10.1093/hebz/9780198724018.001.0001

Ichikawa, J. J., & Steup, M. (2018). The Analysis of Knowledge. In Zalta, E. N. (Ed.), *The Stanford Encyclopedia of Philosophy* (Fall 2018 Edition).

Kahn, J. M., Aulakh, V., & Bosworth, A. (2014). What it will take to achieve the as-yet-unfulfilled promises of health information technology. *Health Affairs*, *33*(4), 536–541.

Kothari, A., Hovanec, N., Hastie, R., & Sibbald, S. (2011). Lessons from the business sector for successful knowledge management in health care: A systematic review. *BMC Health Services Research*, *11*(1), 173. https://bmchealthservres.biomedcentral.com/articles/10.1186/1472-6963-11-173. DOI: 10.1186/1472-6963-11-173

Leonardi, P. M., Neeley, T. R., & Gerber, E. M. (2012). How managers use multiple media: Discrepant events, power, and timing in redundant communication. *Organization Science*, *23*(1), 98–117. DOI: 10.1287/orsc.1110.0638

Moses, N. The development of procedural knowledge in adults engaged in a "tractor-trailer" task. *Cognitive Development*, Volume 9, Issue 1, 1994, Pages 103-130, ISSN 0885-2014, DOI: 10.1016/0885-2014(94)90021-3

Nonaka, I., & Takeuchi, H. (1995). *The Knowledge-Creating Company: How Japanese Companies Create the Dynamics of Innovation*. Oxford University Press. DOI: 10.1093/oso/9780195092691.001.0001

Nonaka, I., & Toyama, R. (2003). The knowledge-creating theory revisited: Knowledge creation as a synthesizing process. *Knowledge Management Research and Practice*, *1*(1), 2–10. DOI: 10.1057/palgrave.kmrp.8500001

Parikh, R., & Renero, A. (2017). Justified true belief: plato, gettier, and turing. In *Philosophical Explorations of the Legacy of Alan Turing: Turing 100* (pp. 93–102). Springer International Publishing. DOI: 10.1007/978-3-319-53280-6_4

Popadiuk, S., & Choo, C. W. (2006). Innovation and knowledge creation: How are these concepts related? *International Journal of Information Management*, *26*(4), 302–312. DOI: 10.1016/j.ijinfomgt.2006.03.011

Stanley, J., & Williamson, T. (2001). Knowing How. *The Journal of Philosophy*, *98*(8), 411–444. DOI: 10.2307/2678403

Star, J. R. (2005). Reconceptualizing Procedural Knowledge. *Journal for Research in Mathematics Education JRME*, *36*(5), 404–411. Retrieved June 19, 2025, from. DOI: 10.2307/30034943

Von Lubitz, D., & Wickramasinghe, N. (2006). Creating germane knowledge in dynamic environments. *International Journal of Innovation and Learning*, *3*(3), 326–347. DOI: 10.1504/IJIL.2006.009226

Von Lubitz, D., & Wickramasinghe, N. (2006a). Networkcentric healthcare: Applying the tools, techniques and strategies of knowledge management to create superior healthcare operations. *International Journal of Electronic Healthcare*, *2*(4), 415–429. DOI: 10.1504/IJEH.2006.010429

Wang, Z., Wang, N., Cao, J., & Ye, X. (2018). The impact of intellectual capital – knowledge management strategy fit on firm performance. *Management Decision*, *56*(9), 1865–1884.

Wenger, E., McDermott, R., & Snyder, W. M. (2011). *Cultivating communities of practice: A guide to managing knowledge*. Harvard Business Review Press.

Wickramasinghe, N. (2006a). Knowledge creation: A meta-framework. *International Journal of Innovation and Learning*, *3*(5), 558–573. DOI: 10.1504/IJIL.2006.010489

Wickramasinghe, N. (2006b). Fostering knowledge assets in healthcare with the KMI model. *International Journal of Management and Enterprise Development*, *4*(1), 52–65. DOI: 10.1504/IJMED.2007.011455

Wickramasinghe, N., & Davison, G. (2004). Making explicit the implicit knowledge assets in healthcare: The case of multidisciplinary teams in care and cure environments. *Health Care Management Science*, *7*(3), 185–195. DOI: 10.1023/B:HCMS.0000039381.02400.49

Wickramasinghe, N., & Schaffer, J. (2010). *Realizing Value Driven Patient centric Healthcare through Technology*. IBM Center for The Business of Government.

Zipperer, L. (2016). Knowledge Management in Healthcare. Routledge. https://books.google.com/books/about/Knowledge_Management_in_Healthcare.html?id=gmUGDAAAQBAJ

Chapter 3
Creating Knowledge and Decision Making

ABSTRACT

Healthcare is fundamentally concerned with decision making. A large part of clinicians' roles involves making decisions around patient care, treatments, discharge and prevention. To do this effectively and efficiently they need complete and correct data and information but also and most importantly germane knowledge. Decision making itself is also not a simple construct and typically is thought of a s a continuum from structured or simple decisions eg; does the patient have an elevated temperature or not, to semi-structured decisions which are moderately complex such as which cough medicine is best, to unstructured decisions which ae extremely complex like in cancer care. In all types of decision-making knowledge is essential but for unstructured decision making the knowledge required is often more reliant on the extraction of multispectral data.

INTRODUCTION

Knowledge creation and decision-making are important in the development of organizations, societies, and economies (Wickramasinghe and Schaffer, 2010; Nonaka & Toyama, 2003). Decision-making is based on the quality of knowledge available that demands systematic generation, accumulation, and utilization. Knowledge management (KM) has a critical role of supporting this process by embracing the human and technological aspects. A socio-technical approach acknowledges that knowledge is not solely a technical construct but a social one that is shaped by the culture, mental processes, and social activity of an organization. This section examines the relationship between knowledge creation and decision making with specific focus on duality of knowledge, knowledge management frameworks, and socio-algorithmic perspectives of knowledge creation.

DOI: 10.4018/979-8-3373-1877-6.ch003

THE SOCIO-TECHNICAL PERSPECTIVE FOR KNOWLEDGE MANAGEMENT

The socio-technical perspective of KM posits that knowledge creation occurs within social systems and technical systems. Organizations are complex systems of human capital that draw upon individual and collective experience, knowledge sharing, and information technologies for creation and utilization of knowledge (Nonaka & Takeuchi, 2019; Nonaka & Toyama, 2003). This perspective is in line with the proposition that KM is not just about technology but is also about understanding the social processes that underpin the creation, storage and use of knowledge.

Theoretical Foundations of Socio-Technical Systems (STS) in KM

Socio-technical systems (STS) theory emphasizes the importance of managing social and technological subsystems for availability and flow of knowledge (Baxter & Sommerville, 2011). The roots of STS can be traced back to the Tavistock Institute in Great Britain in the mid Twentieth Century where observational studies established that the productivity of mining could not only be improved by mechanical technological factors but also by social factors which regulated interpersonal relationships (Trist and Bamforth, 1951). In the same way, today's KM initiatives must integrate the use of technology with the human processes at a workplace.

Knowledge workers in today's world make decisions employing the use of technology, specifically digital platforms, artificial intelligence, and big data analysis (Wickramasinghe, 2006). But there is no denying that human interpretation and contextual understanding cannot be replaced (Davenport & Prusak, 1998). Some of the technologies that support knowledge acquisition and knowledge accessibility include; enterprise knowledge management systems (KMS), artificial intelligence for decision support systems, and machine learning algorithms. However, while used independently, these technologies may generate ineffective or irrelevant information that does not relate to organizational goals (Newell et al., 2009).

The Socio-Technical aspect of KM

The social aspect of KM refers to the people, groups, and culture within an organization that directly impact knowledge management. Expert human knowledge is an important part of the knowledge management system because people bring their practical knowledge, which is defined as the knowledge gained through experience and practice (Polanyi, 2009). Culture that supports knowledge management involves encouraging of employees to learn together, through mentoring and knowledge sharing.

Culture is a critical success factor in knowledge management within organizations. An organisational culture that supports knowledge sharing includes values such as trust, willingness to communicate and share ideas, and knowledge without any repercussions that may deter employees from contributing their knowledge (Dalkir, 2013). On the other hand, a competitive or a departmentalized culture could hamper knowledge sharing, which will prove costly in terms of time and resources due to reinventing the wheel. Managers should also ensure that knowledge is shared across organizational units and levels.

Technological Subsystem: Digital Tools and Knowledge Infrastructure

The technological subsystem includes digital media, databases, artificial intelligence, and information systems that store and process knowledge. The KMS is used in organizations to acquire the tacit knowledge that is captured in explicit knowledge which includes documents, manuals, and best practices (Alavi & Leidner, 2001). AI tools improve KM as they help analyze data, recognize trends, and make forecasts (Becerra-Fernandez & Sabherwal, 2014).

However, one has to understand that technology by itself is not a solution to KM problems. These systems' effectiveness is determined by the ease of use, relevance to the organization's objectives and incorporation into working practices. For example, an organisation that is adopting a KMS has to make sure that its employees are properly trained on how to use the system and that the KMS is integrated within their working processes. Failure to properly implement knowledge sharing systems, or lack of utilization by users can result in knowledge repositories being turned into digital graveyards.

CHALLENGES AND SOLUTIONS IN SOCIO-TECHNICAL KNOWLEDGE MANAGEMENT

There are certain limitations of a socio-technical approach to KM, even though such approach has many potential benefits over the more traditional technical approach. Some of the key challenges are as follows:

- Resistance to Change: Employees may resist implementing new KM technologies as they may feel that these are likely to replace them, lack adequate technical knowledge or may be skeptical about the benefits offered by such technologies (Markus, 2004). This resistance can be overcome through the use of training programs and leadership support in the process of organizational change management.

- Information overload: Due to the exponential growth in the amount of information that is being created and made available in the digital world, it becomes challenging for the employees to quickly locate the required knowledge. It is possible to address this problem through the application of artificial intelligence-based filtering and recommendation systems.
- Knowledge Silos: This means that when departments or teams work independently, knowledge is not shared and thus cannot be used for organizational learning (Riege, 2005). It is suggested that cross functional collaborative activities like the knowledge sharing forums and interdepartmental projects are central to knowledge integration.
- Inadequate Integration of Technology: Some employees may rely on digital tools to carry out their work without seeking assistance from other employees who possess relevant knowledge. To achieve correct contextualization, it is crucial for organizations to strike a balance between reliance on automation and human decision making.

SOLUTIONS FOR ENHANCING SOCIO-TECHNICAL KM

Based on the literature, the following are the managerial strategies that can be deployed for enhancing knowledge management in socio-technical systems:

- Promoting Knowledge Management: Leadership should encourage the sharing of information and best practices among employees. Encouraging knowledge contribution, informing participants of how knowledge is being used, and providing timely incentives can also help.
- User-oriented KM System Design: The digital tools should be easily accessible and should fit into the workflow of the employees. User feedback and usability testing can enhance system usage.
- The need to share both explicit and tacit knowledge: Organizations should implement a balanced approach and both kinds of knowledge-sharing tools. While there are knowledge repositories, there are also informal conversations, mentoring and knowledge sharing through communities of practice.
- Knowledge Management as an Iterative Process: Knowledge management is a continuous process and hence requires constant learning and adaptation. These include the assessment, upgrading of KM systems and training programs that make organizations relevant to change business environments.

DUALITY AND THE KNOWLEDGE CONSTRUCT

Knowledge creation is the conversion of tacit knowledge into explicit knowledge as defined in Nonaka's (2019) SECI model of knowledge management, which consists of Socialization, Externalization, Combination, and Internalization. Explicit knowledge derived from the collective work and documented information is the knowledge that is obtained from the tacit knowledge base of an organization held by individuals. This dynamic process reveals knowledge as an interplay between the tacit understanding of a human and data structures.

It is crucial to recognise that many forms of knowledge are informal and difficult to express, such as abilities, feeling, and shared context learned from the environment. On the other hand, explicit knowledge is organized, documented, and easily transferred through written or taught media, such as documents, computer databases and other teaching aids (Polanyi, 2009). It is important to note that all these forms of knowledge are interconnected in order to foster innovation and facilitate decision making in organizations. The SECI model delineates this interplay into four modes:

- Socialization – Knowledge transfer is also done through activities like, coaching, apprenticeship and cross functional working. This process makes it possible for individuals in an organization to pass on information without necessarily writing it down.
- Externalization – Tacit knowledge is translated into concepts through language, narratives, and frames. This is a critical stage that is responsible for converting individual learning into shared knowledge.
- Integration – Material from different sources is integrated to generate fresh frameworks, reports and strategies. There is a notable use of technologies and structured data management to achieve this by organizations.
- Internalization – Knowledge transferred to an individual is translated into his or her frame of reference and becomes embedded as tacit knowledge after being used.

The SECI model is enhanced by organizational culture, power relations, and cognitive heuristics (Hislop et al., 2018). Companies exist in a social and political context in which the exchange of knowledge is influenced by bureaucratic decision-making and organizational culture. Power dynamics refer to who has the power over information flow and whose knowledge is considered authoritative in an organizational setting. Leaders in organizations are known to prioritize some knowledge types while relegating others to the background, thus maintaining the status quo.

Besides, it is important to note that cognitive biases have a significant influence on the way knowledge is perceived and decisions made. The bounded rationality model by Gigerenzer and Gaissmaier (2011) shows how decision makers work under cognitive and informational limitations, which require the use of frameworks to reduce bias and improve rationality. Heuristics are the mental shortcuts that are used by the individuals and organizations because of their limited cognitive abilities to process information. Heuristics, as effective as they are, bring in their own biases like:

- Confirmation bias – the inclination to accept favorable evidence and reject unfavorable evidence in support of a preexisting claim.
- Availability heuristic – Probability estimation based on easily retrieved information, with a tendency of being wrong in the assessment of certain risks.
- Anchoring – it is a decision-making bias whereby a person depends on initial information as the reference point when making subsequent decisions.

Awareness of these biases enables management to develop KM solutions that help organisations mitigate cognitive biases and make better decisions. For example, structured deliberation frameworks, diverse decision-making teams, and AI analytics can reduce these biases by providing factual perspectives that counter bias.

THE ROLE OF TECHNOLOGY IN KNOWLEDGE DUALITY

The advancement in digital transformation has shifted the duality of knowledge in paradigms in ways such as artificial intelligence, big data, and knowledge management systems. It also supports the storage and retrieval of explicit knowledge and also the codification and sharing of tacit knowledge through collaborative technology tools.

Machine learning algorithms, for example, can work with large sets of data and find some patterns that might be unnoticeable by human decision-makers. However, these systems also contain some bias from their training data and algorithms which require constant monitoring and improvement. Further, AI-based decision support systems improve the availability of knowledge but cannot mimic human instincts and situational awareness.

KMSs are designed to link the two forms of knowledge and provide employees with tools for sharing their experiences and ideas, as well as accessing formalized knowledge. Examples include:

- Enterprise resource planning (ERP) systems – End-to-end business processes and centralised data access.
- Collaboration software (e.g., Slack, Microsoft Teams) – Applications that allow the exchange of information and solutions in real-time.

- Learning management systems (LMS) – Online tools that provide the means to teach and promote constant learning through knowledge sharing and training.

In this regard, there is still the problem of how knowledge is effectively used even as it is created and stored in these technologies. For tacit and explicit knowledge, it is crucial that the organizations establish cultures that support knowledge sharing, critical thinking, and collaboration across disciplines.

CULTURAL AND ETHICAL CONSIDERATIONS IN KNOWLEDGE MANAGEMENT

Culture influences the generation, transfer, and utilization of knowledge within an organization in a great way. Knowledge sharing foster communication, trust, and cooperation while knowledge management where some individuals or some departments retain information will retard organizational innovation. The following are ways through which organizations can foster knowledge-sharing:

- Incentive structures – Providing employees with motivation for posting to knowledge databases and for teaching others.
- Transparent communication policies – Encouraging openness and accessibility of information across hierarchical levels.
- Diversity and inclusion – The ideas of equality, multiculturalism, and pluralism.

Another factor that cannot be overlooked is that of ethical issues in knowledge management. Concerns like ownership of ideas and information, privacy of the data, and the dissemination of fake news must be addressed to uphold the highest standards of business ethics. For instance, an organization dealing with sensitive information must employ appropriate security measures and standards of ethics to avoid unauthorized access to the data.

Firstly, in an age where there is so much information available, it becomes difficult to differentiate between accurate knowledge and fake knowledge. The availability of large amounts of unverified information on the internet requires efficient means of verification and media literacy programs to increase the share of checked knowledge.

THE FUTURE OF KNOWLEDGE DUALITY

As the process of knowledge creation proceeds, the identification of the roles of tacit and explicit knowledge will remain significant in organizations. Some of the trends that will define this change include:

- Augmented intelligence – The use of artificial intelligence in combination with human talent to improve decision-making.
- Decentralized knowledge sharing – Knowledge sharing has shifted to a decentralized system through the use of blockchain technology.
- Cognitive Computing – Sophisticated AI systems that aim at replicating human thinking hence enhancing knowledge processing and decision making.

Companies that can successfully implement these innovations and strike a proper balance between human instinct and technology, would be well prepared to address more complicated issues and advance innovation on the long-term perspective.

FRAMEWORKS FOR KNOWLEDGE CREATION

There are many frameworks that offer structured methodologies for the creation of knowledge allowing organizations and individuals to capture, disseminate, and utilize knowledge. Tacit and explicit knowledge conversion is a continuous and cyclic process of converting information into explicit knowledge that can be used for innovation, problem solving and decision making. Several theoretical frameworks have been proposed to explain and describe this process, and all of them provide different insights into how knowledge is created, shared, and applied.

Among the most popular models that explain the process of knowledge creation is the SECI model that originates from the works of Nonaka and Takeuchi (2019). This model shows how tacit and explicit knowledge are in a constant flow, divided into four modes: Socialization, Externalization, Combination, and Internalization. According to the SECI model, it is the process of converting these two forms of knowledge that leads to knowledge creation. Socialization entails passing of implicit knowledge through interpersonal contact in the form of apprenticeship or observation. Externalization refers to the articulation of tacit knowledge into explicit concepts, often through documentation or discussion. Integration is a process of knowledge organization and its systematization, which results in the formation of new forms of structured knowledge. Finally, Internalization is the final stage where the acquired knowledge is taken and converted into a practice that is personal and organizational, through the use of experience. This cycle guarantees a constant exchange of knowledge within an organization, which leads to innovation and development.

Another framework is Boisot's (1998) I-Space model that distinguishes between two types of knowledge – codified and tacit – and its distribution across organizations. According to Boisot the processes of knowledge creation and dissemination can be broken down into six possible stages, namely Scanning, Problem Solving, Abstraction, Diffusion, Absorption and Impact. During the Scanning phase, the organization defines what information is appropriate to gather and/or acquire. Of all the steps, the Problem-Solving phase entails evaluating and making sense of this information. Abstracting means making generalizations so as to develop broader knowledge which is transferable across different settings. The Diffusion phase involves disseminating this knowledge to other appropriate stakeholders, while the Absorption phase involves the process by which people make sense of the knowledge in relation to their own worldviews. Lastly, the Impact phase checks the applicability and implications of the newly gained knowledge. The I-Space model underscores the process of knowledge codification and how structured knowledge dissemination is instrumental in getting the right knowledge to the right people at the right time.

Besides these basic models, another model that captures the dynamic nature of knowledge development is the Firestone and McElroy's (2012) Knowledge Life Cycle model. However, in this model, there are only two fundamental cycles: the knowledge production cycle and the knowledge integration cycle. Knowledge creation process refers to the development, testing, and improvement of new knowledge in the form of ideas, solutions, and findings through research, experimentation and problem solving. Once validated, this knowledge flows into the integration phase of knowledge management where it is used, transferred and incorporated into the organizational systems. The Knowledge Life Cycle model also emphasizes feedback and feedback loops, which means that the knowledge is reviewed and updated in light of its application and new findings. Thus, knowledge validation and refinement mechanisms are useful in improving the decision making process as well as promoting a culture of continuous improvement in organizations.

Another theoretical framework that offers another way of looking at knowledge creation is Wiig's (1994) Knowledge Management Cycle that focuses on knowledge construction, application and reproduction. Knowledge management framework suggested by Wiig includes four categories of activities: Building Knowledge, Holding Knowledge, Pooling Knowledge and Applying Knowledge. The Building Knowledge phase focuses on both learning and gaining experience to formulate new knowledge that forms the foundation of the solution. Knowledge retention can be defined as the process of retaining the Knowledge in databases, documents, and people. Pooling knowledge refers to the collection and sharing of knowledge across the various organizational levels. Last but not least, Applying Knowledge explores how knowledge is used to make decisions, generate ideas and solve issues. According to Wiig (1994), one of the major preconditions for knowledge to remain useful and applicable is that the organization continues learning and adapting, thus integrating the particular knowledge into the organizational structure and processes. In this, culture, leadership and knowledge networks within an organization are important determinants of knowledge sharing and innovation. Research has found that knowledge creation is most effective in organizations that encourage information sharing and cooperation across functions and promote diverse knowledge stocks. Through the development of a culture that supports the management and development of knowledge, organizations can fully exploit the value of their intellectual assets and foster sustainable development.

Thus, there are several frameworks that can offer a structured view on how knowledge creation can be approached, and all these frameworks have their merits regarding the generation, sharing, and application of knowledge. The SECI model, I-Space model, KLC model, and the KMC each play a role in explaining how knowledge develops within organizations. Also, the development of digital tools and the acknowledgment of cultural and social contexts enhance our understanding of knowledge-building processes. The use of these frameworks and embracing other related trends will help organizations improve their knowledge management and increase their competitive advantage in today's knowledge based economy.

A SOCIO-ALGORITHMIC APPROACH TO KNOWLEDGE CREATION

Recent developments in artificial intelligence and machine learning have brought forth a socio-algorithmic approach to knowledge construction (Šajeva, 2010). In modern KM frameworks, data analysis algorithms and tools explore structured and unstructured information to recognize patterns and patterns to aid decision-making. AI integration into knowledge management has boosted prediction analysis, knowledge search, and the optimization of organizational IQ (Brynjolfsson & McAfee, 2017). But there is also a set of problems associated with this growing reliance on algorithms in knowledge creation such as the problems in the selection of data, ethical issues, and disconnection from the real world (O'Neil, 2016). However, to harness the power of AI for knowledge creation and yet, at the same time, to avoid the potential harm, AI needs to be embedded in an organisational approach that is consistent with reliability, fairness, and ethical use of AI. The following strategies can be useful in the achievement of this balance:

- Auditing AI-generated insights: Human experts should monitor and verify the AI-generated insights for accuracy and context relevance. Appointing interdisciplinary review committees can help in decision-making processes and increase confidence in AI systems.
- Increasing AI Transparency: Organizations need to focus on the development of explainable AI methods to improve interpretability of the models. SHAP (Shapley Additive Explanations) and LIME (Local Interpretable Model-Agnostic Explanations) can reveal certain information about the behavior of AI models and their decision-making processes.
- Implementing AI best practices: AI best practices, which include fairness, accountability, and inclusiveness, should be incorporated into AI design and deployment. Some of the recommended ethical guidelines include those from the Institute of Electrical and Electronics Engineers (IEEE) and the European Commission IEEE (2022).
- Real-time Monitoring and Auditing: The AI models should be monitored and audited on a consistent basis to identify any bias, mistakes, and decline in efficiency. The feedback loop that incorporates the opinion of the users and real data can be used to improve the model and increase the accuracy of the artificial intelligence systems in the future.

The socio-algorithmic approach to the generation of knowledge can therefore be regarded as a revolutionary paradigm in the production, organization and application of knowledge. By combining AI with analytics with the help of professionals, organizations can use machine learning while keeping ethics and contextual factors in check. However, to address challenges like algorithm bias, ethical issues, or reliance on automation, it is essential to be proactive about fairness and transparency and include human oversight. Through the implementation of Responsible AI, organizations can capture the full benefits of AI in knowledge creation in this age of digital transformation and enhance organizational intelligence.

CONCLUSION

This chapter has discussed the knowledge creation and decision-making within the socio-technical and socio-algorithmic frameworks to understand interactions between humans and technology. With the rise of AI and ML in the organization and knowledge generation, it becomes crucial to comprehend both the dual nature of knowledge, explicit and tacit, as well as the framework within the creation process. AI is a powerful tool that can process large amounts of data, find patterns, and make predictions, but for this it has to incorporate human logic, values, and context.

The socio-technical approach focuses on both the technology and the people involved in creating knowledge, as it is acknowledged that knowledge is not simply produced by formula but is influenced by social, cultural, and other factors. Such models such as Nonaka and Takeuchi, 1995 SECI model explain how knowledge flows between the explicit and tacit forms via the four processes of socialization, externalization, combination, and internalization. It improves these processes through the use of artificial intelligence particularly in data classification, summarization and predictive analysis. However, as the production of knowledge is taken over algorithmically, it may become rather divorced from human understanding and therefore may be more prone to bias or ethically unsound conclusions.

The socio-algorithmic perspective develops this analysis further by examining the part played by AI in knowledge management. Algorithmic systems help in decision-making by processing large amounts of information within a short period but come with issues like bias, lack of transparency, and dependency on algorithms. AI bias can be caused by the dataset used while training the model, historical issues, or design issues with algorithms. If left unchecked, these biases can lead to discrimination, reduce the inclusion of diverse viewpoints, and decrease confidence in AI outcomes. Future research should focus on methods for bias identification, fairness evaluation, and interpretability to make AI-generated knowledge more ethical. Efforts should also be made to understand how ethical AI can be implemented in the knowledge generation process and how this process can be made less biased with higher levels of interpretability. To foster the responsible development of AI, we need to have engagements from data scientists, ethicists, policymakers, and other domain specialists. It has been observed that integrating technology with the human element can create trustworthy, context-sensitive, and ethical decision-making mechanisms.

In other words, the process of knowledge creation in the digital age requires a socio-algorithmic perspective that considers issues of fairness, transparency, and the nature of AI's incorporation. In the current and future world of AI, collective action is necessary to ensure that this technology improves rather than erodes the decision-making ability and knowledge creation of the human mind.

REFERENCES

Alavi, M., & Leidner, D. E. (2001). Review: Knowledge management and knowledge management systems: Conceptual foundations and research issues. *Management Information Systems Quarterly*, *25*(1), 107–136. DOI: 10.2307/3250961

Baxter, G., & Sommerville, I. (2011). Socio-technical systems: From design methods to systems engineering. *Interacting with Computers*, *23*(1), 4–17. DOI: 10.1016/j.intcom.2010.07.003

Becerra-Fernandez, I., & Sabherwal, R. (2014). *Knowledge management: Systems and processes*. Routledge.

Boisot, M. (1998). *Knowledge assets: Securing competitive advantage in the information economy*. Oxford University Press.

Brynjolfsson, E., & McAfee, A. (2017). *Machine, platform, crowd: Harnessing our digital future*. W. W. Norton & Company.

Dalkir, K. (2013). *Knowledge management in theory and practice*. MIT Press. DOI: 10.4324/9780080547367

Davenport, T. H., & Prusak, L. (1998). *Working knowledge: How organizations manage what they know*. Harvard Business Press.

Firestone, J. M., & McElroy, M. W. (2012). *Key issues in the new knowledge management*. Routledge. DOI: 10.4324/9780080495972

Gigerenzer, G., & Gaissmaier, W. (2011). Heuristic decision making. *Annual Review of Psychology*, *62*(1), 451–482. DOI: 10.1146/annurev-psych-120709-145346 PMID: 21126183

Hislop, D., Bosua, R., & Helms, R. (2018). *Knowledge management in organizations: A critical introduction*. Oxford University Press. DOI: 10.1093/hebz/9780198724018.001.0001

IEEE. (2022). *Ethically aligned design: A vision for prioritizing human well-being with autonomous and intelligent systems*. Institute of Electrical and Electronics Engineers. https://ethicsinaction.ieee.org/

Markus, M. L. (2004). Technochange management: Using IT to drive organizational change. *Journal of Information Technology*, *19*(1), 4–20. DOI: 10.1057/palgrave.jit.2000002

Newell, S., Robertson, M., Scarbrough, H., & Swan, J. (2009). *Managing knowledge work and innovation* (2nd ed.). Palgrave Macmillan. DOI: 10.1007/978-0-230-36641-1

Nonaka, I., & Takeuchi, H. (1995). *The knowledge-creating company: How Japanese companies create the dynamics of innovation.* Oxford University Press. DOI: 10.1093/oso/9780195092691.001.0001

Nonaka, I., & Takeuchi, H. (2019). *The wise company: How companies create continuous innovation.* Oxford University Press.

Nonaka, I., & Toyama, R. (2003). The knowledge-creating theory revisited: Knowledge creation as a synthesizing process. *Knowledge Management Research and Practice*, *1*(1), 2–10. DOI: 10.1057/palgrave.kmrp.8500001

O'Neil, C. (2016). *Weapons of math destruction: How big data increases inequality and threatens democracy.* Crown Publishing Group.

Polanyi, M. (2009). *The tacit dimension.* University of Chicago Press.

Riege, A. (2005). Three-dozen knowledge-sharing barriers managers must consider. *Journal of Knowledge Management*, *9*(3), 18–35. DOI: 10.1108/13673270510602746

Šajeva, S. (2010). *The analysis of key elements of socio-technical knowledge management system.* Economics & Management.

Trist, E. L., & Bamforth, K. W. (1951). Some social and psychological consequences of the longwall method of coal-getting: An examination of the psychological situation and defences of a work group in relation to the social structure and technological content of the work system. *Human Relations*, *4*(1), 3–38. DOI: 10.1177/001872675100400101

Wickramasinghe, N. (2006). Fostering knowledge assets in healthcare with the KMI model. *International Journal of Management and Enterprise Development*, *4*(1), 52–65. DOI: 10.1504/IJMED.2007.011455

Wickramasinghe, N., & Schaffer, J. (2010). *Realizing Value Driven Patient centric Healthcare through Technology*. IBM Center for The Business of Government.

Wiig, K. M. (1994). *Knowledge management foundations: Thinking about thinking—how people and organizations create, represent, and use knowledge*. Schema Press.

Chapter 4
Managing Knowledge Complexity

ABSTRACT

Today we live in a data intensive economy. Thus, turning data assets into actionable knowledge is essential. Healthcare is one if not the most data rich industries and thus managing knowledge complexity in healthcare organizations is a central issue and a strategic necessity. This chapter serves to unpack critical issues in this regard. Specifically the chapter examines critical people, process and technology issues that must be considered for knowledge to be successfully extracted from raw data elements.

INTRODUCTION

The current corporate landscape has seen organizations grappling with knowledge complexity. Such complexity is solidified by forces like the ever-quickening pace of technological change, the globalization of markets, and the growing specialization of knowledge workers. Here, effective knowledge management (KM) is vital for organizations to manage this complexity, and turn it into a strategic advantage.

In this chapter, we examine the knowledge complexity and propose an organizational model for KM that can help organizations manage this knowledge complexity effectively. It delves into the distinction between being prepared and being ready, the value of training to cope with knowledge complexity, and the use of the OODA (Observe, Orient, Decide, Act) loop, developed by Colonel John Boyd, in KM practice. The chapter ends with a summary of key points and implications for practice.

DOI: 10.4018/979-8-3373-1877-6.ch004

AN ORGANIZATIONAL MODEL FOR KM

Knowledge complexity must be managed following a structured approach in line with the strategic needs of an organization. The actual implementation of knowledge management (KM) could be tailored based on organizational models, as it provides a practical context for how knowledge is developed, shared, and used within an organization. It must be a living model that is responsive to the changing nature of knowledge while offering specific advice regarding the management of knowledge. By not having a clear, structured model, organizations expose themselves to inefficiencies, knowledge silos, and most importantly, the inability to leverage their intellectual capital as a source of competitive advantages. This oriented conceptualization model of KM is formed of some key elements which included knowledge creation, knowledge storage and retrieval, knowledge sharing, knowledge application, and knowledge evaluation. The combination of all aspects is essential for the effective management and use of knowledge to meet the organizational objective.

Knowledge Creation: knowledge creation means crafting new knowledge through research, innovation, and collaboration. To enable knowledge creation, organizations need to build a creative and experimental culture. According to Nonaka & Takeuchi (1995), a knowledge-creating company is one where tacit knowledge (know-how, know-what) is converted to explicit knowledge (know-that) through processes known as socialization, externalization, combination and internalization. For instance, companies such as Google and 3M allow employees to devote part of their work hours to innovative projects, resulting in the development of new information and products. Creating a culture that fosters knowledge requires the commitment of leaders, open and transparent forms of communication, and the ability to take risks.

Knowledge Storage and Retrievals is yet another important aspect of the organizational model. Systems that can help store and access knowledge when needed makes it effective KM Practices. Databases, document management systems, and knowledge repositories are examples of such platform. Alavi and Leidner (2001) demonstrate that the possibility of using information technology to store and retrieve knowledge was only possible if those systems were user-friendly and became an integral part of employees' routine functions. For example, companies such as IBM and Siemens have deployed advanced knowledge management systems which allow employees to find and retrieve information across departments and geographies. But the effectiveness of these systems is contingent upon their design, ease of use and compatibility with the requirements of an organization.

Knowledge Sharing is key to make sure the knowledge gets dispersed across the organization. Knowledge is siloed without effective sharing, which hampers its ability to influence and hence the need to share it via communities of practice.

Wenger (1998) describes communities of practice as groups of people with common interests or expertise who collaborate while solving problems and sharing information. For instance, communities of practice have been used successfully by organizations like NASA and the World Bank to facilitate the sharing of best practices and lessons learned across projects. Sharing knowledge is not just about having a platform or system in place, but also about creating an environment in which employees feel safe and encouraged to share their expertise.

Knowledge Application is the final phase of KM. It uses knowledge to enhance organizational performance, solve problems and add value. According to Davenport and Prusak (1998), the real measure of KM success is how knowledge is used in practice and leads to real improvements. This means systems that allow people to embed knowledge into their work, including decision-making frameworks, best practice guidelines, performance metrics, and the like. The application of knowledge of lean manufacturing principles to improve efficiency has resulted in the development of the Toyota Production System (TPS), which is commonly referred to as the Just-in-Time production system. Knowledge application necessitates agreement between what an employee wants and what an organization needs along with a system of rewards that encourages knowledge application, known as user acceptance or IT acceptance.

The ultimate block of the organizational model is Knowledge Evaluation. There is the need to review the effectiveness of KM practices on regular basis to make sure the practices are not missing the content of KM. This includes evaluating the effect of knowledge on performance of the organization, examining gaps and areas of opportunity, and making adjustments accordingly. Grant (1996) argues that at this stage organizations have to assess the extent of knowledge integration i.e., the way different kinds of knowledge are put together in order to generate value. For example, companies like Procter & Gamble measures things like rates of innovation, time-to-market, and customer satisfaction. This is often used as an indicator of knowledge evaluation which is a commitment to learn and improve from past performance, from those classical failures and successes to enhance KM practices for future implementation.

Implementing an organizational model for KM requires a strategic approach that involves leadership commitment, resource allocation, and continuous improvement. Leaders drive KM initiatives and set the tone for a culture that recognizes knowledge as a strategic asset. This means having the appropriate infrastructure in place to support KM activities: this will include allocating resources, including funding, technology, and training. According to Zack (1999), alignment of KM with organizational strategy focuses on knowledge as the key driver of competitive advantage. This has put KM in the center stage of strategic planning and in enabling innovation and growth for organizations such as Apple and Amazon. It is also critical that organizations establish metrics to assess the effectiveness of their KM practices, and use these metrics to fuel continuous improvement. To do this, there is the need to be making data an integral part of decision-making process and be able to pivot as conditions change.

PREPARED VERSUS READY

Preparedness and Readiness in knowledge management are two distinct dimensions critical to managing knowledge complexity. Preparedness is how organizations proactively prepare for uncertainties and solve for challenges that relate to knowledge. It involves building knowledge centers, training professionals, and building robust KM processes. On the opposite end of the spectrum is readiness, which describes an organization's ability to react to unexpected knowledge-related challenges as they emerge (Weick & Sutcliffe, 2001). While preparedness is predominantly about insight and design, readiness is about diversity and flexibility at the time of an ambiguity. Organizations need both concepts to manage complexities of knowledge in a dynamic and persistent environment.

The Importance of Preparedness

Preparedness is important for effective KM because organizations can step up to identify the likely areas of problems and plan for contingencies. For example, organizations that invest in comprehensive knowledge repositories and training programs are better equipped to handle disruptions such as the loss of key employees or the introduction of new technologies. Knowledge repositories are a collection of information available in a single place and help in keeping the knowledge intact when the employee leaves the organization or retires. On the other hand training programs provided for employees make sure they have the skills and knowledge to perform their roles effectively and are less likely to leave knowledge gaps (Teece, Pisano, & Shuen, 1997).

Practical preparedness also means that organizations can be more agile to respond when things change in the external environment (for example, changes in market demand and regulatory changes). This allows organizations to prepare contingency plans and reformulate their plans accordingly. Here is an example of how it can happen: A company that foresees changes in its industry regulations can update its compliance processes and train employees on new requirements to ensure they are minimally affected during that change and they can continue their operations. Therefore, preparedness not only helps reduce the risks, but it also opens the door for opportunities for the organization to surpass others by being ahead of the competition (Teece, Pisano, & Shuen, 1997).

However, preparation is not only about groundwork for the external context but also for making an internal ecosystem for knowledge understanding and cooperation. So, the focus is on formal KM processes, such as knowledge audits and knowledge mapping, to identify critical knowledge assets and retrieve for it to the right person. It also entails establishing a culture of shared learning, allowing up -to -date knowledge to be shared to all employees as needed. Therefore, it is important for organizational leaders to create a culture that values knowledge sharing and collaboration (Nonaka & Takeuchi, 1995).

The Importance of Readiness

While preparation is very important, readiness is just as important in order to manage knowledge complexity. Readiness is an organization's capability to respond to unforeseen challenges, which goes beyond what can be prepared for before and to some extent during a crisis. In an age with unpredictability and uncertainty, organizations need to reconstitute their knowledge and rewiring their mental models to adapt the dynamic complexity of real world so as to cease it with real knowledge to make imperative decisions. In order to achieve readiness, there needs to be an evolutionary culture of agility, flexibility and continuous learning with empowered employees, who can think on their feet and take initiative (Eisenhardt & Martin, 2000).

One of the key aspects of readiness is the ability to leverage tacit knowledge, which is the implicit, experience-based knowledge that employees possess. This contrasts with explicit knowledge, which can be codified, stored in repositories, and transferred through training programs (Nonaka, 1994). The promulgation of tacit knowledge through informal networks and communities of practice allows organizations to work with the knowledge that is already present in the firm, enabling employees exchange their individual expertise as well as specialize and learn from one another. This not only enhances the organization's collective knowledge but also enables employees to adapt to new situations and solve problems creatively (Polanyi, 1966).

Part of readiness is being able to make quick decisions when certainty is lacking. Employing this requires a decentralized decision-making structure, where employees at every level have the power to take action based on their knowledge and expertise. Under these types of environments, decisions are not made only by top management but are shared actions dispersed throughout the organization so that the organization can react quicker and more effectively to surprises. In fast-paced and high-stake industries like tech and healthcare, this also is critical (Eisenhardt & Martin, 2000).

In addition to the capability to learn from experience, there are many other key facets of readiness. Well-prepared organizations have systems to gather lessons learned from past experiences and apply them to future challenges. This includes, for example, conducting post-mortem analyses after major projects or crises, documenting best practices, and sharing these insights across the organization. This aligns with the idea that organizations must continually build on KM practices to better respond to future problem challenges and is an example of learning from experience (Garvin, 1993).

On the Balance of Preparedness and Readiness

KM effectively needs to balance preparation with readiness. Though preparedness builds a solid base of knowledge resources and processes, readiness is what ensures that organizations can pivot amid unforeseen challenges and capitalize on new opportunities (Senge, 2006). Strong preparedness requires strategic planning. This includes understanding knowledge asset needs, creating KM processes, and allocating time and resources for training and infrastructure. But it isn't just about the need for readiness that must be considered in strategic planning. Rather a culture of nimbleness, agility and the ability for employees to take roots and adapt as situations demand should also be factored. This means creating a culture of continuous learning, in which employees are encouraged to try new things, take chances, and learn from their failures (Senge, 2006).

Learning is a core tenet of preparation as well as readiness. This is both a measure for keeping employees up to speed on skills and knowledge necessary to execute their jobs, which plays into preparedness, and ensuring the organization has the right people in the right positions. This also encourages a culture of adaptation and innovation, thus increasing readiness. This way, organizations can help develop their employees, not just to face today's challenges but tomorrows too, by investing in continuous learning (Garvin, 1993).

Another key ingredient balancing preparedness and readiness is adaptive leadership. Leaders contribute significantly to determining the organisation's KM orientation. These leaders need to be advocates for KM programs, allocate resources to support them, and work to create a culture where sharing knowledge and bringing forth new ideas are celebrated. Simultaneously, leaders need to be flexible and willing to change as they should be ready to adjust what they previously thought and how they did things when new information presents itself. With adaptive leadership, leaders can influence their employees to be both prepared and ready, thus engendering a culture that is proactive and responsive (Heifetz & Laurie, 1997).

ROLE OF TRAINING

Training is vital in the operation of managing knowledge complexity inside organizations. It enables the staff with adequate skills and knowledge to execute their responsibilities effectively, contributing significantly towards the organizational pool of knowledge. An effective training program allows employees to obtain new information and to develop expertise, as well as problem solving skills. Training promotes a culture of ongoing learning to help organizations prepare for changes in their industries. In fact, this is especially critical for knowledge-intensive contexts, where managing and applying information effectively can determine the success of the organization (Garvin, 1993).

Knowledge transfer is a key component of training in knowledge management. Customary organizations develop extensive storage of both explicit and tacit knowledge, and training is a mean to bridge this gap of knowledge to employees. Employee training serves as a catalyst for career growth by enabling employees to build competencies that allow them to handle the complexities involved in their work environment (Baldwin & Ford, 1988). Additionally, even the best training can only take an organization so far, and maintaining continuity in the knowledge-sharing process becomes impossible when employees leave — by ensuring that training is an ongoing process, an organization can prevent the knowledge loss often associated with employee turnover. In fast-paced business environments, knowledge complexity grows when new technologies, regulations, and industry standards come on-stream. A strong training program implementation is crucial for organizations to avoid being left behind their competitors. And at the same time employees are being trained to stay versed with the latest trends and innovations which eventually helps the organization to perform better than its competitors.

Types of Training

Companies can adopt various kinds of training to address knowledge complexity successfully. Efficiency is one of the main benefits of organizing different types of training. Depending on their goals, workforce makeup, and industry needs, organizations should choose the training methods that would work best for them.

Formal Training Programs: These programs are designed to help employees learn specific information and abilities. Formal training comes in the form of things such as workshops, seminars, certifications and online courses. These programs, often taught by industry experts or internal trainers, are geared towards one-sided information transfer. Formal training programs work best in ensuring that employees learn both knowledge and best practices that are consistent with organizational goals (Noe, 2017). Formal training provides structured learning for the employees to be competent in a controlled environment to inculcate their professional abilities. Hence, organizations spend on formal training programs to comply with regulations within the industry, develop tactical skills as well as make better decision-making (Optimum Control, 2018). Many of these programs also include assessments to measure learning outcomes, helping to ensure that participants have effectively understood key concepts.

On-the-Job Training: As opposed to formal training, on-the-job training is the primary focus for this type. Workers perform on-the-job tasks while simultaneously learning new skills. It is this method that enables the person to convert learned knowledge into hands-on experiences in real situations, furthering the understanding via practical experiences. According to Polanyi (1966), tacit knowledge cannot always be codified or conveyed through formalized training programs, which makes on-the-job learning an especially powerful tool. Tacit knowledge encompasses the intuition, problem-solving, and situational awareness an employee learns over the course of their work experience. Thirdly, employees learn through doing with on the job training, they acquire the experience of how the organization functions, they improve in the area they work. Additionally, this type of training encourages mentorship and peer learning, as employees tend to learn through the guidance of experienced colleagues while accomplishing tasks. The culture of continuous learning fostered by organizations that prioritize on-the-job training enables employees to remain adaptable and evolve to solve complex challenges.

Mentoring and Coaching: Mentoring and coaching is another way of training in knowledge management. These approaches emphasize individualized interactions between seasoned employees and less experienced. Mentoring and coaching programs encourage tacit knowledge transfer, where employees can learn from seasoned practitioners who have years of industry experience (Kram, 1985). Traditional training has a топ-to-bottom and linear structure, while mentoring and coaching enable a more fluid way of learning where both parties – mentor/coach and mentored/coached person – learn и grow together. Employees are provided with support tailored to their unique needs, allowing them to navigate complex work environments with greater confidence. In addition, professional development is one of the advantages of mentoring as it presents opportunities and professional advancement and skill improvement. Exchanging knowledge through mentoring within organizations enhances employee engagement, knowledge retention, and succession planning. Employees are guided through constructive feedback and performance evaluations in a coaching environment to help them improve their skills and face hurdles. Seeking mentorship not only provides guidance, but helps the institution retain know-how and continue to thrive with competent personnel.

Communities of Practice: Organizations can foster knowledge sharing through communities of practice in addition to formal training programs, on-the-job training and mentoring. These are informal groups of employees with common interests or specialties. These communities are platforms for collaborative learning where employees share best practices and exchange insights on industry trends (Wenger, 1998). Communities of practice are especially useful in dealing with knowledge complexity by creating an atmosphere of peer-to-peer learning. Colleagues share insight and experience from the real world, and employees are able to learn new problem-solving methods. The organisation which support community of practices foster a culture of innovation and continuous improvement. These communities also provide an important vehicle for disseminating knowledge; employees can record what they learn and add to shared bodies of knowledge. Organizations can also use this approach to better adapt to changing market conditions and emerging challenges.

ASSESSMENT OF TRAINING IMPACT

To ensure that training programs are effective in managing knowledge complexity, organizations must implement evaluation mechanisms. Evaluating the effectiveness of training programs helps organizations pinpoint strengths and weaknesses. Surveys, evaluations, assessments, and performance metrics are some of the methods to gain insight into the effectiveness of training. The evaluation process gives an idea of whether employees have learn and apply the new knowledge in their jobs (Kirkpatrick, 1994).

A popular evaluation framework is the Kirkpatrick Model, which reports evaluation across four levels: reaction, learning, behavior, and results. Level 1: Reaction The reaction level evaluates the participants' sentiments toward the training experience, asking for their feedback on elements like relevance, interest, and delivery quality. The learning level refers to knowledge gained through tests and quizzes. The behavior level analyzes the modification of personnel behavior towards the application of new skills learned and adopted, and the results level evaluates how much the training process has affected the organization's performance. Using this framework, organizations can appreciate what is achieved by training and use data to inform decisions aimed at improving future training activities.

Key performance indicators (KPIs) are another way to assess training effectiveness. Before and after such courses too employees' performance can be measured by organizations to see the improvements to their productivity, efficiency and knowledge retention. Moreover, follow-up assessments enable organizations to measure the lasting effect of training on employee proficiency and job satisfaction.' This constant assessment is key for maintaining training course relevance, as knowledge-intensive industries evolve quickly, such that skills required change over time, train-outs become redundant and need to be agreed in advance with business goals.

Moreover, organizations can introduce learning feedback loops, where employees provide feedback on the training effectiveness and recommend opportunities for improvement. So, by promoting a culture of open feedback, organizations can improve their training programs to address the skills gap as well as the industry trends. The purpose of training evaluation is to improve how knowledge is managed so that employees have the proficiency mobile in the face of complexity to lead the organization to greater heights.

OODA LOOP - BOYD

Overview of the OODA Loop

Colonel John Boyd devised a decision-making framework known as the OODA loop (it is made up of four stages: Observe, Orient, Decide, Act). This framework is especially important in the case of knowledge management (KM) in general (grasping the complexity of knowledge in uncertain and moving environments). By appropriately utilizing the OODA loop, organizations can become better equipped to adapt to changes, make better decisions, and stay ahead of the competition (Boyd, 1987).

Applying the OODA Loop to KM

Notice

The first step of the OODA process is to take cues of the world surrounding us. In the realm of KM, this pertains to keeping an eye on shifts in the commercial landscape, technological developments, and changes in the regulatory environment. Organizations should also pay attention to internal flows of knowledge to find gaps and opportunities for improvement. Organizations must set up strong data collection systems, use powerful analytics tools, and keep communication open that allows information to flow freely between departments to observe effectively. The speed of collecting and processing relevant information can make the difference between an organizations being agile to respond to emerging challenges (Boyd, 1987)

Orient

The second step in the OODA loop is to interpret the information gained in the observation phase. Hence, organizations have to review that data, detect patterns, and interpret the information in line with their strategic goals. By orienting raw data, organizations are able to harness the data to benefit the organization, making validate this a very important piece in dealing with knowledge complexity. Organizations will need to build analytical capabilities and use more advanced techniques including machine learning and artificial intelligence to scale their capabilities to process large amounts of information. Moreover, organizations need to address the impact of cognitive bias and incorporate diverse perspectives in decision-making processes to avoid judgment errors. At the same time, a more robust orientation stage enhances an organization's ability to move through uncertainty with more judgment and clarity (Boyd, 1987)

Decide

The third step is to make choices based on the analysis done in the orientation step. Within KM, this includes decisions around knowledge creation, sharing, and application. Decision-making that works well is a mix of gut-feel and analytical thinking with a touch of flexibility. Structured decision-making frameworks that enable accountability and transparency must be implemented by organizations. In addition, decisions for KM should derive from organizational goals and indirectly align with overarching strategic aims. It is important to create decision-making culture for organizations that accepts trials and learning from failures and it will help in addressing organisations to revise their path and improve knowledge intensive process. Making timely decisions based on knowledge management is crucial to ensuring that they are relevant or not (Boyd, 1987).

Act

The last stage is implementing the decisions from the previous stage. When it comes to KM, this means deploying knowledge management projects like training programs, knowledge-sharing frameworks or innovation projects. Knowledge and know-how alone will not improve performance; Action is therefore vital. Successful implementation relies on effective leadership, the ability to manage change, and the active engagement of employees. Without establishing clear performance metrics, organizations are unable to determine whether the things that they are doing are effective or not, essentially reaching the desired outcomes. By continuously assessing the impact of their initiatives, organizations can identify areas for improvement and make necessary adjustments to enhance their KM strategies (Boyd, 1987).

CONTINUOUS IMPROVEMENT THROUGH THE OODA LOOP

The OODA loop is not a one-time process but a continuous cycle of observation, orientation, decision-making, and action. Organizations must continuously iterate through the OODA loop to adapt to changing circumstances and improve their KM practices. This requires a culture of continuous learning, agility, and adaptability. Organizations that embrace the iterative nature of the OODA loop can foster innovation and enhance their resilience in an increasingly complex and uncertain business environment. By leveraging feedback loops and promoting knowledge-sharing behaviors, organizations can ensure that the OODA loop remains an integral part of their knowledge management strategy (Boyd, 1987).

CONCLUSION

Managing knowledge complexity is a critical challenge for organizations operating in dynamic and uncertain environments. This chapter has presented an organizational model for KM that provides a framework to address this knowledge complexity. This framework is composed of integral components such as knowledge creation, storage and retrieval, sharing, application, and evaluation. These components come together to enable knowledge management and provide a boost to organizational performance. Other aspects this chapter has reviewed include the need for differentiation between being prepared and being ready, the significance of training in knowledge complexity management, and the use of OODA loop in KM practices.

An important element of good KM is finding the balance between preparedness and readiness. Preparedness requires you to put systems and processes in place to manage knowledge, while readiness is about applying knowledge to situations as they happen. Both are necessary for organizations to be able to pivot and adapt to change. This includes regularly updating and enhancing KM initiatives, along with investing in employee training and development, which will help them remain adaptable and capable of navigating intricate knowledge landscapes.

Since a knowledge-based society is characterized by the complexity of knowledge, training is a crucial pillar for KM, as it provides workers with the skills and competencies that enable them to deal with such complexity. Explicit and tacit knowledge are built through various forms of training (formal programmes, on-the-job training and stuff like mentoring and communities of practice). Conducting a tailored training program builds a knowledge-sharing and learning culture promoting the KM process across the organization.

One widely used decision-making framework in KM is the OODA loop, which stands for Observe, Orient, Decide, and Act. It enables organizations to handle, evaluate information, determine its relevance, make decision and take actions in a systematic manner. This iterative process ensures that organizations adapt to new challenges and adjust their KM strategies based on feedback from the field.

REFERENCES

Alavi, M., & Leidner, D. E. (2001). Knowledge management and knowledge management systems: Conceptual foundations and research issues. *Management Information Systems Quarterly*, *25*(1), 107–136. DOI: 10.2307/3250961

Boyd, J. R. (1987). *A discourse on winning and losing*. Maxwell Air Force Base, AL: Air University Press.

Davenport, T. H., & Prusak, L. (1998). *Working knowledge: How organizations manage what they know*. Harvard Business School Press.

Eisenhardt, K. M., & Martin, J. A. (2000). Dynamic capabilities: What are they? *Strategic Management Journal*, *21*(10-11), 1105–1121. DOI: 10.1002/1097-0266(200010/11)21:10/11<1105::AID-SMJ133>3.0.CO;2-E

Garvin, D. A. (1993). Building a learning organization. *Harvard Business Review*, *71*(4), 78–91. PMID: 10127041

Grant, R. M. (1996). Prospering in dynamically-competitive environments: Organizational capability as knowledge integration. *Organization Science*, *7*(4), 375–387. DOI: 10.1287/orsc.7.4.375

Kirkpatrick, D. L. (1994). *Evaluating training programs: The four levels*. Berrett-Koehler Publishers.

Kram, K. E. (1985). *Mentoring at work: Developmental relationships in organizational life*. Scott, Foresman.

Noe, R. A. (2017). *Employee training and development*. McGraw-Hill Education.

Nonaka, I. (1994). A dynamic theory of organizational knowledge creation. *Organization Science*, *5*(1), 14–37. DOI: 10.1287/orsc.5.1.14

Nonaka, I., & Takeuchi, H. (1995). *The knowledge-creating company: How Japanese companies create the dynamics of innovation*. Oxford University Press. DOI: 10.1093/oso/9780195092691.001.0001

Polanyi, M. (1966). *The tacit dimension*. University of Chicago Press.

Senge, P. M. (2006). *The fifth discipline: The art and practice of the learning organization*. Broadway Business.

Teece, D. J., Pisano, G., & Shuen, A. (1997). Dynamic capabilities and strategic management. *Strategic Management Journal*, *18*(7), 509–533. DOI: 10.1002/(SICI)1097-0266(199708)18:7<509::AID-SMJ882>3.0.CO;2-Z

Weick, K. E., & Sutcliffe, K. M. (2001). *Managing the unexpected: Assuring high performance in an age of complexity*. Jossey-Bass.

Wenger, E. (1998). *Communities of practice: Learning, meaning, and identity*. Cambridge University Press. DOI: 10.1017/CBO9780511803932

Zack, M. H. (1999). Developing a knowledge strategy. *California Management Review*, *41*(3), 125–145. DOI: 10.2307/41166000

Chapter 5
The Organization's Human Infrastructure

ABSTRACT

Healthcare professionals play a very key role in care delivery. One way to understand this is to frame them as knowledge workers, people who have expertise, both tacit and explicit knowledge. By sharing and utilizing this knowledge in an optimal, efficient and effective fashion it is also possible to provide high quality, high value patient centered care. Thus the organisations human infrastructure plays a critical role in ensuring successful operations. This in turn leads to the delivery of high value, patient centred care any time anywhere for any one. Strategies to do so are presented in this chapter.

INTRODUCTION

In a knowledge focus context, the human infrastructure within an organization is its greatest strength (Lee et al., 2006). Unlike material or technological assets, human capital is the total number of employees skills, expertise and intellectual efforts, especially those of knowledge workers, which are important for innovation and staying competitive (Drucker, 2012). Organizations need to be able to capture, retain, and utilize knowledge for long term success. However, managing human infrastructure involves dealing with a variety of challenges including organizational culture, effective leadership, change management, and structural adaptation (Nonaka, 2009).

This chapter explores the fundamental issues in human infrastructure where knowledge workers are highlighted, effective knowledge capture techniques discussed, and organizational elements, such as culture, leadership, and change adaptation covered in detail. By understanding these aspects, companies can better leverage their employees for efficiency, creative thinking and lasting growth.

DOI: 10.4018/979-8-3373-1877-6.ch005

KNOWLEDGE WORKERS

Drucker (2012) explained "knowledge worker" concept for the first time as a name for people whose primary benefit is knowledge. Whereas those who work with their bodies use their theoretical and analytical skills to confront difficult issues and create value, knowledge workers use their theoretical and analytical skills to address complex problems and create value. Knowledge worker jobs cover a wide variety of positions including scientists, engineers, IT specialists, healthcare professionals, educators, analysts and managers who are all expected to have proficiency and critical thinking skills in an ever-changing environment (Lee et al., 2006). The growing relevance of intellectual capital means that knowledge workers are on the frontline of innovation, and productivity improvements, and the creation of enduring competitive advantage for organizations. The value of the knowledge workers are in the quality and innovation of their work and not in the input of their time, according to Davenport (2005).

Definition and Role

The essence of the definition is the understanding that knowledge workers operate in cognitive contexts (Lee et al., 2006). Their employment is dependent on their ability to use specific knowledge for appropriate situations; they tend to transform complex theories into tangible models, methods, or outputs. Their work calls for more than mere performance of repetitive duties (Davenport, 2005); They are commonly required to address ambiguity, make decisions in volatile environments and facilitate and/or share insights across teams and corporate structure. These workers are typically engaged in work that requires interpretation, synthesis, and inventive intellectual effort. Data scientists are an integral part of the organizations, where they collect, organize and, most importantly, analyze data that will be used to facilitate informed business decisions. As another example, policy analysts examine existing laws, detect trends in public behavior, and offer strategic counsel for governments on future policy changes. Their analytical approach is critical for enabling advancement, innovation, and flexibility in enterprises and even government agencies (Lee et al., 2006).

Characteristics of Knowledge Workers

The features that characterize knowledge workers can be used to differentiate them from the traditional staff and understanding such characteristics is important in managing them effectively (Drucker, 2012). Autonomy is the most significant attribute of knowledge workers. When it comes to organizing their workload, overcoming challenges, and deciding course of action, Independence highly matters for knowledge workers. Their productivity falls in authoritarian environments, but they thrive in settings where independent thinking and creative action are possible. Autonomy according to Pink (2011) has a fundamental role to play in motivating knowledge workers. The provision of autonomy on how they manage their work gives the knowledge worker the ability to come up with not only original solutions but more productive results.

Furthermore, an extensive dedication to lifelong learning is another prominent characteristic. The dynamic nature of knowledge-intensive sectors calls for the need to update the knowledge and skills of the workers to ensure that they remain effective. Motivated by an intense desire to learn, these workers always seek to learn, invest in their professional development, and co-operate with colleagues to stay ahead in their professions. Based on Senge (1990), the foundation of "learning organization" lies in fostering continuous growth of individuals. In such a situation, knowledge workers are provided with improved opportunities, and thus more likely to help achieve organizational goals.

While knowledge workers are of great value, there are special obstacles in the management of such workers (Abualoush et al., 2018). Although routine workers are mainly tasked with the implementation of predetermined tasks, knowledge workers tend to undertake tasks that require the formation of new approaches, products, or methods. As Florida wrote in his 2002 work, the creative class is distinguished by their work of innovating, problem solving and progressing technology economically (Florida, 2002). Knowledge workers, who can offer new perspectives, challenge traditional approaches and take unproven alternatives, are a valuable asset for organizations working in dynamic markets.

Challenges in Managing Knowledge Workers

Although knowledge workers deliver enormous benefits, there are some unique management challenges that go beyond the norms of the ordinary managerial practices (Carleton, 2011). Firms that ignore such challenges risk losing good employees and preventing new ideas from developing. A persistent problem is the rate at which the knowledge workers quit their jobs. The specialization skills and the high industry demand for such skills often give knowledge workers numerous job opportunities and they tend to often move from one organization to another. Individuals seeking these positions often base their decisions on more than paying them. In most cases, the need to move on is driven by a desire to undertake more purposeful assignments, or to experience more growth potential, or better well-being at work and home. According to Cappelli (2008), the main way in which these individuals are retained is through intrinsic motivators i.e., recognition, autonomy and further professional development as opposed to mere salary increments or bonuses.

A further concern is the need to ensure that employees never slack off (Carleton, 2011). Knowledge workers are more concerned about where they find themselves working than what they may do to achieve that position. As Amabile and Kramer (2011) indicate, it is important to keep knowledge workers motivated if they are to keep making progress in an important work. When employees are involved in projects aligned with personal goals and have a purpose, they become more productive, more creative, and more dedicated to remaining with the company. In place of this, if workers feel their tasks are unchallenging or not aligned with overall goals, disinterest and a need to quit arise.

The other more subtle problem is the act of knowledge hoarding (Carleton, 2011). In some work environments, knowledge workers can hold information closely to maintain job stability or place themselves more advantageously compared to peers. This is such behavior which is known as knowledge hoarding, and which is a major barrier to collaboration, innovation, and communication. This tendency is common in organizations that do not promote collaboration or create confidence among its staff members (Hislop 2013). If knowledge workers believe revealing information will weaken their role or security, then they are not likely to cooperate with team goals. This situation shows how important transparent, inclusive, and mutually supporting organizational cultures are.

Organizations need to adopt a leadership style that embraces support as opposed to directive control in order to manage these obstacles. The managers of knowledge workers should move from direct supervision to more of a coaching role, supporting, and not controlling the ways work gets done. They should encourage teams' cooperation, acknowledge creative contribution, and develop the mechanism for the constructive feedback. Move performance management from strict numerical benchmarks to a more qualitative evaluation of creativity, problem solving, collaboration and continuous improvement.

Organizations can benefit from implementing systems that will be used to gather, archive and share information across departmental lines. By embracing these tools, organizations reduce the probability that valuable information gets locked in a single employee and also protect against the disruption that arises when key members exit. Promoting building of cross-functional teams and the act of peer-to-peer mentoring can ensure that expertise is more evenly spread and applied.

In the end, the value of knowledge workers to the success of today's organizations cannot be overemphasized. Their ability to innovate, their constant desire for learning and their intellectual contributions make them important players in fast changing and high competitive fields. But then, considering their unique characteristics, the management style should be adapted too. Organisations need to embrace new methods, departing from strict command hierarchies to embracing autonomy, professional growth, and cooperative work cultures. Understanding the distinctive requirements and dreams of knowledge workers helps businesses to develop initiatives that attract top talent and foster long-term engagement and production. Knowledge is becoming increasingly a very important asset in the economy, and the management of knowledge workers is a key success factor for organizations.

CAPTURING KNOWLEDGE FROM KNOWLEDGE WORKERS

In the current knowledge economy, many firms lose their competitive edge less from physical capital than from the ability to tap the knowledge of their workforce (Carleton, 2011). The very nature of the knowledge and experience that the knowledge workers bring to the table makes them excellent repositories of intellectual capital. However, a great part of the value of this knowledge depends on an organization's ability to capture, distribute and apply it in the company. If organizations fail to systematically collect and disseminate this knowledge, they may lose ground to valuable insights, particularly when the key staff leave to work in other positions. With this in mind, collection of knowledge from staff posits as essential not only because of strategic reasons, but as well very crucial to sustain innovation and the growth of the organization.

Explicit vs. Tacit Knowledge

Knowing the challenges and strategies in capturing knowledge involves the identification of two major categories of knowledge: explicit knowledge and tacit knowledge. This differentiation, one of the crucial insights of Nonaka (2009) work is a mainstay in the current approaches to managing organizational knowledge.

Explicit knowledge refers to content that is straightforward to describe in words, write and make available to others for their use. Numerical data, procedural standards, technical instructions, research reports, and various types of written information are examples. Organizational repositories such as books, databases, spreadsheets, and wikis are populated with explicit knowledge that can be easily shared and duplicated within an organization. Being articulated and stored, explicit knowledge enables widespread use and consistent application across the company. For example, implementing SOPs or creating training materials enables companies to share knowledge on a large scale, guaranteeing that all the locations and teams do the work up to the same standard.

However, understanding tacit knowledge is much more difficult as it is a repository of the on-the-ground experience, wisdom, and instincts developed by people as they live their lives (Nonaka, 2009). Tacit knowledge may be difficult to express in writing or formal words because it may develop from inner thought processes or may be based on a specific background of a person. As a seasoned software developer, one may rely on subconscious problem-solving skills after trying and trying again, yet it is not easy to explain to someone who is not familiar with the situation exactly how he did it. Similarly, a project manager may have an in-depth, subtle understanding of how to motivate a team, manage the concerns of stakeholders, or identify early warning signs of a project's potential pitfalls – often based on subtle intuition, rather than written guidelines. This difference presents a critical challenge for organisations to harmonise. While it is easy to capture and manage explicit knowledge, tacit knowledge is more important because it drives innovation, good decision-making and creative responses to challenges which are important in delivering results. In this regard, any holistic attempt at knowledge management should take into account both types and should focus on tacit knowledge discovery and sharing strategies in meaningful ways.

KNOWLEDGE MANAGEMENT SYSTEMS (KMS)

As organizations try to manage and transmit both explicit and tacit knowledge, they increasingly turn on Knowledge Management Systems (KMS) (Maier & Hadrich, 2011). Knowledge Management Systems merge several technologies and collaborative methods to collect, classify and distribute knowledge within the organization.

Organizations are using the practice of documenting explicit knowledge as a common practice. Such efforts include establishing of clear standard operating procedures, compilation of training materials, production of technical documents, and establishing internal platforms such as wikis or intranet portals. According to Alavi and Leidner (2001), such methods help to retain and share information that is available for use by existing and future employees. When an employee leaves the organization, their recorded procedures and accumulated insights can be used to support new employees which will reduce knowledge attrition and make onboarding times shorter.

But in order to obtain tacit knowledge, organizations should pay attention to developing relationships and experiential learning. A very effective thing to do is to create mentorship programs. In these programs, senior employees and new hires are paired, so that knowledge is shared through observation, supervised work, and storytelling. Argote and Ingram (2000) contend that mentored relations are essential for imparting hidden expertise and undocumented organizational routines. The senior employees as mentors can demonstrate their problem solving style, contextual thinking and strategic choices in decision making effectively filling the knowledge gaps that may not be apparent in standard training materials.

The formation of Communities of Practice (CoPs) is another very effective method to promote knowledge sharing (Meyerhoff & Strycharz, 2013). CoP refer to a group of people who assemble around a common profession or interest that constantly engage in dialogue to enhance skills and overcome issues (Wenger, 1999). The community groups are natural forums through which knowledge workers can share experiences, get advice and work together to enhance their knowledge. For example, software engineers might hold a CoP to talk about new technological advancements, trade solutions to complex bugs, or offer constructive critique of one another's code during collaborative sessions. Consequently, participants often share tacit knowledge that would otherwise be lost if those interactions did not take place.

From the above, KMS therefore go beyond databases and digital tools. They depend on social networks which promote teamwork and the learning of new skills. Organizations that employ a blend of technology resources and relationships with people are better able to capture and utilize a broad array of knowledge assets.

Challenges in Knowledge Capture

Companies tend to face major obstacles when trying to manage and capture knowledge (Novak et al., 2020). Organizations often face two key problems, reluctance to share information and the weight of handling too much data.

Specialized and competitive workers are sometimes the most reluctant to share their knowledge. Davenport and Prusak (1998) point out that withholding knowledge is not unusual because employees are worried about safeguarding their personal value, they fear that their contribution will not be noticed, or that their information will be misused. Sometimes the organisational structure can subtly discourage sharing by emphasising individual successes rather than encouraging teamwork. The second key issue is information overload. Even though documentation is essential for the preservation of explicit knowledge, too much documentation or unstructured information can negatively affect productivity and frustrate employees searching for helpful content. In order to meet this barrier, an organizational culture shift is necessary whereby it is made to feel safe, honored and rewarded by the company when sharing knowledge. There is a great extra obstacle of the abundance of information. This can be caused by too much documentation or an absence of organization. This is inefficient and frustrating as employees try to access information. Wilson (2002) argues that too many documents may prevent access if search capabilities are poor or classification systems are lacking. Employees may abandon formal systems for more effective methods, such as informal networks instead of official resources.

The second major challenge is information overload. Documentation of explicit knowledge is important but too much documentation or jumbled information can actually be counterproductive: Decreased productivity and increased dissatisfaction of workers who try to find the necessary information. As Wilson (2002) explains, the sheer volume of documentation can overwhelm users when the information systems do not have simple search mechanisms or a coherent categorization. When the access to meaningful information becomes complicated because of excess or repetitive information, staff may abandon formal systems and resort to unstructured channels or reinstating old ways.

Therefore, knowledge management is only effective when it provides a balance between collecting knowledge and storage and retrieval mechanisms. Strategic approach is necessary for organizations to handle and address these challenges. Such actions involve the use of friendly technologies that facilitate efficient searching as well as designating staff members to manage content and make sure that users contribute to designing the organization's knowledge frameworks. When knowledge-sharing is embedded in individual and team metrics, it is normalized rather than exceptional throughout the organization.

For organizations that want to maintain their competitive position and long term viability, initiative in extracting valuable knowledge from the workforce has to be a priority. While the explicit knowledge is simple to document and store, tacit knowledge, which is not uncommonly that most valuable, requires intentional and relational processes such as mentorship and Communities of Practice. The effective implementation of Knowledge Management Systems is based on the synergy of technology and interpersonal interaction. Still, companies have to overcome well-known barriers like the reluctance to share information and the risk of being overwhelmed, because these phenomena can hinder the success of any well-thought knowledge initiatives. Developing a trust base, encouraging cooperative efforts, and structuring knowledge in a way that allows enterprises to leverage individual capabilities into collective wisdom supporting enterprise wide success.

ORGANIZATIONAL CONSIDERATIONS

In the rapidly changing business world, firms have to adjust continually to the rapid advancement of technology, changing market trends, changing demands of the consumers, and increasing global competition (Novak et al., 2020).

Therefore, organizations that are agile and can react strategically have become critical in sustaining and moving them further in business. Change adaptation has become compulsory, not optional. For organizations to remain competitive, they need to constantly evaluate their internal systems, structures, and procedures due to external changes. At the heart of the organizational transformation is the art of effective change management, a process that is systematic in nature, and that aims at facilitating the successful implementation of new procedures, technologies, or cultural changes within a company.

The organizational transformation requires more than the introduction of new software or policy changes. Its main purpose is to help staff members through the process of transformation, to keep them engaged, minimize setbacks, and contribute to long-term growth. The discipline of change management concerns itself with integrating both the procedural changes and people related factors of organizational transitions. The process involves understanding the emotional reactions of employees, evaluating the need for structural change in the organization and effective leadership in handling the change. If organizational change is not handled with the use of an intentional strategy and implementation, initiatives may fail whereby resources are lost, employees are opposed and the pace of change is slowed down. The application of the tested models of Kotter (2012) and Lewin (1947), in particular, is of crucial importance for leading the organizations through change in a logical and efficient way.

CHANGE MANAGEMENT

The 8-step Change model created by John Kotter and formed in 1996 is a basic framework for managing changes within an organization that many people consider (Lauer, 2010). Kotter's framework presents a sequential set of actions to create momentum, collect stakeholder support, and facilitate a long-term change implementation (Kotter 2012). It is not only a procedural list; rather, it is a single scheme that addresses the rational and emotional aspects of change in organizations. The model starts with the critical task of creating urgency within the organization. Any change initiative needs to have a compelling justification of why change is necessary to be presented first. Causes for change may include worsening outcomes of operations, changing demands from the market, or new digital technologies. Employees and management will not be motivated if there is no sense of urgency. The goal is to create motivation among members of the team and point out the risks of maintaining the current state. Leaders can help create a sense of urgency by providing data, analyzing market trends and showing competitive challenges in order to make it clear that action needs to be swift.

On the basis of the urgent foundation, the next step is to establish an influential coalition that will drive the change effort forward. No single leader can drive or maintain a major change on his own. A group of stakeholders with a variety of opinions must work together to promote and advance the initiative across the organization. These people for the coalition must be influential, credible and capable of inspiring people. Leaders in the guiding coalition are not only expected to develop the change strategy; they need to champion it, build trust and overcome resistance.

Once urgency has been set, the organization needs to come up with a vision and a strategy of change. An open vision explains the reason for the change and the expected outcomes for employees. The strategy that follows outlines the processes and ways of achieving the envisaged goals. A strong vision is simple, clear and emotionally powerful. Ambiguity in the vision can result in change efforts losing focus and stakeholder commitment reducing.

Once a vision is created, Kotter highlights the importance of communicating that vision regularly. Leaders are encouraged to communicate with staff in frequent, uniform, and varied channels of communication, including meetings, emails, workshops, and daily interactions. When leaders do their best to work in line with the vision, their commitment is made clear and their behavior is congruent. This approach minimizes uncertainty, promotes common understanding, and tells employees that the change is taken seriously and will be lasting. This, in turn, provides spaces for input and thus makes employees feel like part of the process.

Kotter's model specifies that step five empowers staff members to lead sweeping action. This means eliminating all barriers that prevent the initiations of change. Obstacles may be out dated processes, cultural fears like fear of failure, or managerial barriers like unsupportive supervisors. To empower the employees is to give them the tools, training, and organizational support needed to make significant participation in the change possible. This stage means celebrating those who are called change pioneers and encouraging those who are still adjusting.

Then organizations should focus on short term successes. Such early wins are fundamental in maintaining the forward momentum and for demonstrating that organizational transformation is effective. They show that goals can be achieved and they encourage sceptics to become believers. Organizations should purposefully plan and proclaim easy victories to keep the belief alive that change brings real results. Emphasizing the role played by both employers and teams during these initial stages can go a long way in boosting morale and motivating.

At this phase, the organization tries to formalize progress and create even more significant shifts. Organisations should not relax even when initial wins are won. Rather, they provide a foundation to disrupt the status quo and even bigger-scale transformations. Major activities in the phase include reflection of what worked well, revising plans and overcoming remaining barriers that are part and parcel of the organizational systems and operations. Kotter observes that change initiatives fail at this point these days because businesses tend to believe that they are done before they really are. Actual change requires constant work and further progress.

Kotter goes further to recommend including innovative approaches into the organizational culture. If change is to be lasting, it needs to be embedded within the culture of the organization. It refers to the integration of new practices and attitudes into everyday routines, assessment process, education, and leadership training. The manner in which work is carried out in day to day life is called culture and the integration of change in the unwritten rules of the day is the foundation for long term achievement. Leaders should constantly walk the talk and associate rewards with the new ways of doing things to achieve cultural alignment.

Resistance to Change

Change is sure to occur, but it is often challenged by resistance on individual and collective levels (Lauer, 2010). Resistance can take many forms, ranging from suspicion, to indifference, disengagement and even actual opposition. Understanding what fuels resistance is essential to managing it properly. Some of the common sources of resistance include apprehension about the unknown, fear of giving up control and concern over job security as well as frustration at how changes have been treated in the past.

Lewin's (1947) early contributions to change management theory remain a critical guide for addressing this challenge (ibid). Three-stage process was suggested by Kurt Lewin, who said: unfreeze, change, and refreeze. During unfreezing, organizations need to intentionally instigate a state of instability and a feeling that the current approach is not working any longer to prepare for meaningful change. The process of communication is very important at this stage as it ensures that the employees understand the reasons behind the changes, and what they are to gain. During unfreezing, the leaders should ensure that they support and comfort employees by answering their concerns honestly.

In change stage, introduction of revised processes, approaches and mindsets occurs. This step can be particularly difficult because it involves the breaking of old patterns and getting used to novelty of processes. Making sure that employees are engaged, they receive clear information, and are encouraged by supportive leaders, contributes to a less impactful transition during this time. Empowerment, training, and quick results in this stage will help to build confidence and reduce resistance.

During refreezing, there is an attempt to formalize and stabilize the new norm. It involves the embedding of the changes by revising the policies, revising job descriptions, introducing new reward systems, and establishing new performance standards. Marking milestones and assimilation of change into daily routines make it a permanent part of the company.

Transparency, employee engagement, and clear communication are found to be important components in Kotter's and Lewin's frameworks. Involving the employees in the change process and informing them about the importance of this process is a powerful way to make them accept new initiatives. Further, fostering a trusting environment and frequent communication will reduce uncertainty and improve organization's adaptability.

Although complex, the success of effective organizational change is essential to survive in today's dynamic markets. With the rapid changes in outside conditions, companies should aim to be nimble, proactive, and adaptable. However, Kotter's 8-Step and Lewin's Three-Stage Models offer important guidelines to help in change, in responding to organizational imperatives as well as in considering the employees. Essentially, successful change efforts explain that changing organizations is essentially human, not only process or infrastructure. Open communication, empowering staff, and the incorporation of new routines in organizational culture are key for the development of sustainable achievements. In volatile settings, the capacity for change management and implementation is not optional, but essential.

ORGANIZATIONAL CULTURE AND STRUCTURE

As we see the emergence of the knowledge-based industries, the organizational culture and structure begin to play a critical role in determining the ability of the entity to access and utilize its intellectual capital. Organizational culture and structure go way beyond administrative issues; Their power is pervasive, (Janićijević, 2013) shaping everything from people's communication patterns to their decision making and organization's collective knowledge management (Ichsan et al., 2023). These forces are critical in shaping the way learning flows around the organization, from one team or department to another and within the business hierarchy. As knowledge workers and innovation become more important for organizations to remain competitive, the need to examine the way culture and structure influences organizational outcomes has never been more important.

Culture's Impact on Knowledge Sharing

Organizational culture has a strong influence on the sharing of knowledge in an organization (Janićijević, 2013). The cultural ethos and practices of an organization support or hinder the flow of knowledge in an organization. Schein (2010) asserts that organizational culture defines the way in which employees relate to each other, their relationships and how they evaluate the worth of working together. If trust, openness and respect are the building blocks of a collaborative culture, employees will naturally share knowledge which feeds organizational learning and innovation.

This practice generally results in the development of dynamic informal linkages, constant ideas exchange and a clear view of what the organization is striving to achieve.

The Google collaborative culture shows how knowledge sharing can be enhanced greatly. Google's inclusive and open culture promotes collaboration between workers in different departments and investments in initiatives that are not part of their main responsibilities and thus, encourages innovation. According to Laszlo Bock (2015) in Work Rules!, Google's in-house procedures and organizational ethos promote the idea of psychological safety amongst the employees who are therefore encouraged to share raw ideas and innovative thinking. Such a setup promotes innovation by removing barriers between teams and creating room for varied views exchange.

On the other hand, if organizations focus on personal achievement rather than cooperation, they may restrict the flow of knowledge between team members. In competitive workplaces, the employees can withhold knowledge to ensure he/she is safe or advance his/her career. Anxiety to lose status or recognition frequently prevents the team members from working together and impedes the process of innovative ideas' generation. By concentrating on individual awards instead of team successes, organizations can inadvertently encourage secrecy and can reduce overall team collective intelligence. To overcome these cultural barriers, companies have to establish learning and cooperative values within their organizations.

Structural Adaptability

The process of information handling and decision making in an organization is closely linked to its general structure and culture (Janićijević, 2013). Formal organizational structure defines how roles, tasks, and authority are spread within the organization. Structures define the flow of tasks within the organization, who reports to whom, the flow of information, and ultimately, the ability of the organization to respond to change. Henry Mintzberg (1979) disclosed that the difference between hierarchical and flat organizational designs has significant impacts on the speed of decision-making, the level of autonomy that the staff have, and the flexibility of the organization.

A hierarchical structure entails a chain of management levels and the existence of express authority hierarchies. Such structures guarantee order and role specificity, but they may also inhibit flexibility and restrict the transmission of critical information from the employees to the top management. In very bureaucratic environments, top management tends to make too centralized decisions, and this may lead to inefficiency and reduced communication between the leaders and the front lines. This rigidity can hinder innovation and reduce the ability to respond to new problems quickly. Flat structures, in turn, streamline management hierarchies and empower employees with more decision making authority. In such settings, the organizations are more responsive, adaptable and team supportive. Flat organizational structures allow free communication and often lead to the staff becoming more involved in their work. With more empowerment sense, workers will also become more motivated and thus more opportunities for proactive knowledge exchange and experimentation will arise.

Many organizations choose between hybrid or matrix structures in order to find a balance between the traditional hierarchy of the classic management and the flexibility of flat structures. In a matrix arrangement, the employees will be found in both functional and project roles hence will be answerable to more than one supervisor depending on the tasks at hand. According to Galbraith (2014), hybrid frameworks enable both maintenance of functional expertise and fast adaptation to the changing market demands or opportunities for innovation. In spite of the added complexity in terms of coordination, organizations that use such structures can often thrive in areas that always demand innovation and teamwork in diverse teams.

MANAGEMENT AND LEADERSHIP

Similar to culture and structure, leadership has a great impact on the way knowledge is managed, and innovation prospers (Silva, 2016). The effective leaders are critical to both shaping the priorities of the strategy and shaping the environment that drives and engages employees. The leadership approach applied in a company can stimulate staff to provide their insights and develop professionally, or discourage motivation and decrease individual participation.

Transformational Leadership

Transformational tendencies in leadership are characterized by inspiring followers with a vivid and ambitious picture, and intellectually pushing them and providing them with personalized encouragement (Silva, 2016). Transformational leaders are highly successful in inspiring creativity and maintaining high employee commitment (Bass, 2006). They encourage team members to look at situations objectively, question underlying beliefs, and test alternative approaches to resolve problems. To knowledge and intellectual abilities-led organizations, transformational leadership is a key competitive advantage in driving success as it focuses on intrinsic motivation and intellectual engagement.

Transformational leadership upholds a standard by providing an example of practices that foster fairness, honest interaction, and equality among all team members. The convergence of personal ambitions with organizational goals enhances a team identity, compelling employees to participate emotionally and intellectually.

According to transformational leadership, knowledge workers need to cooperate, share their skills, and seek original solutions to the organizations' problems.

Servant Leadership

Servant leadership emerges as a model that is getting acclaimed as a means of promoting wholesome effects among knowledge workers (Silva, 2016). The term 'servant leadership' coined by Robert Greenleaf in 1977, emphasizes supporting and facilitative roles of the leaders as opposed to the command. Leadership based on the servant model is focused on serving employees, pursuing improvement in their personal and professional growth. In turn of concentrating on people, servant leaders foster trust, loyalty, and high engagement in team members.

The effectiveness of servant leadership in settings centered on knowledge is in creating a workplace where employees feel empowered through valuing their efforts, listening and understanding their views and supporting them with the necessary tools. By working on the development of their team, the leaders motivate the employees to respond with dedication, enthusiasm, and spirit of information exchange. Servant leaders are able to build cohesive communities, where the spirit of collaboration and mutual support for learning and innovation is enhanced, by eliminating the conventional hierarchies.

Servant leadership is correlated with increased employee satisfaction and lower employees' turnover rates. Moreover, the servant leadership focuses on nurturing rather than exercising authority, which makes it easier to instill leadership skills internally – a crucial strategy in maintaining knowledge and expertise for longer horizons.

CONCLUSION

The relationship between organizational culture, structure, and leadership has overall critical importance to the ability of a firm to manage its human resources and encourage innovation. A cooperative and transparent environment promotes knowledge exchange while inflexible or adversarial cultures do the opposite. Similarly, organizations that are introducing flat or hybrid systems frequently achieve a greater level of agility and enhanced communication in the workforce, distinguishing them from other organizations in the rapidly changing markets that focus on innovation.

Good leadership is an integral part of how the ecosystem operates. The leadership styles that focus on inspiration, support and personal development i.e. transformational and servant leadership are notably effective at motivating knowledge workers and developing a vibrant organizational culture. These leadership approaches are complementary to the basic values of autonomy, creativity and continuous learning as defining knowledge work.

An organization's strength is supported by its talented workforce and the management practices. The dynamic advancement of this infrastructure is mainly founded on knowledge workers through their combined knowledge, creative thinking, and collective spirit that push organizational growth. To the greatest extent, companies need to utilize leadership techniques that promote flexibility, trust, and continuous learning for people and collaborative groups. Such initiatives contribute towards retaining key staff in the organization, while allowing sustained excellence in the midst of changing global business realities.

REFERENCES

Abualoush, S., Bataineh, K., & Alrowwad, A. A. (2018). The Role of Knowledge Management Process and Intellectual Capital as Intermediary Variables between Knowledge Management Infrastructure and Organization Performance. Interdisciplinary. *Journal of Information, Knowledge, and Management, 13*, 279–309. DOI: 10.28945/4088

Alavi, M., & Leidner, D. E. (2001). Knowledge management and knowledge management systems: Conceptual foundations and research issues. *Management Information Systems Quarterly, 25*(1), 107–136. DOI: 10.2307/3250961

Amabile, T., & Kramer, S. (2011). *The progress principle: Using small wins to ignite joy, engagement, and creativity at work.* Harvard Business Press.

Bass, B. M. (2006). *Transformational leadership.* Lawrence Elabaum Associating. DOI: 10.4324/9781410617095

Bock, L. (2015). *Work rules!: Insights from inside Google that will transform how you live and lead.* Twelve.

Cappelli, P. (2008). *Talent on Demand.* Harvard Business School Publishing.

Carleton, K. (2011). How to motivate and retain knowledge workers in organizations: A review of the literature.

Davenport, T. H. (2005). *Thinking for a living: how to get better performances and results from knowledge workers.* Harvard Business Press.

Drucker, P. (2012). *Management challenges for the 21st century.* Routledge. DOI: 10.4324/9780080942384

Florida, R. (2002). *The rise of the creative class* (Vol. 9). Basic books.

Ichsan, M., Sadeli, J., Jerahmeel, G., & Yesica, Y. (2023). The role of project management office (PMO) manager: A qualitative case study in Indonesia. *Cogent Business & Management, 10*(2), 2210359. Advance online publication. DOI: 10.1080/23311975.2023.2210359

Janićijević, N. (2013). The mutual impact of organizational culture and structure. *Economic annals, 58*(198), 35-60.

Kotter, J. P. (2012). *Leading change.* Harvard business press.

Lauer, T. (2010). *Change management.* Springer Berlin Heidelberg. DOI: 10.1007/978-3-642-04340-6

Lee, C. P., Dourish, P., & Mark, G. (2006). The human infrastructure of cyberinfrastructure. *Conference on Computer Supported Cooperative Work*, 483–492. DOI: 10.1145/1180875.1180950

Maier, R., & Hadrich, T. (2011). Knowledge management systems. In *Encyclopedia of Knowledge Management* (2nd ed., pp. 779–790). IGI Global.

Meyerhoff, M., & Strycharz, A. (2013). Communities of practice. *The handbook of language variation and change*, 428-447.

Nonaka, I. (2009). The knowledge-creating company. In *The economic impact of knowledge* (pp. 175–187). Routledge.

Novak, A., Breznik, K., & Natek, S. (2020). How leaders can initiate knowledge management in organizations: Role of leadership style in building knowledge infrastructure. *Human Systems Management*, 39(1), 37-50. (Original work published 2020)DOI: 10.3233/HSM-190596

Pink, D. H. (2011). *Drive: The surprising truth about what motivates us*. penguin.

Schein, E. H. (2010). *Organizational Culture and Leadership*. Jossey-Bass.

Senge, P. M. (1990). *The fifth discipline: The art of organizational learning systems*. Bantam Doubleday Dell Publishing Group.

Silva, A. (2016). What is leadership?. *Journal of business studies quarterly, 8*(1), 1.

Wenger, E. (1999). *Communities of practice: Learning, meaning, and identity*. Cambridge university press.

Wilson, T. D. (2002). The nonsense of knowledge management. *Information Research, 8*(1), 8–1.

Section 2
The Technology Perspective

Chapter 6
Digital Twins in Healthcare Discovering Complex Biological Rules for Achieving Precision and Personalization in Healthcare

ABSTRACT

The pursuit of precision and personalization in healthcare has led to investigating computational means to model sophisticated patient-related aspects. Extending this line of enquiry, this study investigates the application of Digital Twins (DTs) to model patients—i.e., to create virtual representations of patients that integrate multi-omic and clinical data—to uncover complex biological rules that can inform treatment planning. As a clinical case study, we focus on Triple-Negative Breast Cancer (TNBC). We examine how DTs can be used to enhance the planning of PD-1/PD-L1 inhibitor therapies for TNBS. PD-1/PD-L1 is a class of immune checkpoint inhibitors with variable patient responses for TNBC. Our approach aims to computationally identify patterns of statistical congruence in cellular information that may predict therapeutic outcomes, thereby enabling clinicians to determine in advance which patients are most likely to benefit from specific immunotherapies.

DOI: 10.4018/979-8-3373-1877-6.ch006

INTRODUCTION

Precision and Personalization in Healthcare

Precision and personalization of care is an emerging area in healthcare that aims to tailor treatments to the unique characteristics and needs of each patient (Goetz & Schork, 2018; Wang & Wang, 2023). This involves considering a wide range of factors—such as genetics and other biological markers, comorbidities, lifestyle, diet, and environment—to move beyond the traditional "one-size-fits-all" model. The goal is to deliver the right treatment to the right patient at the right time, improving effectiveness while reducing side effects.

Achieving this level of personalization relies heavily on computational methods (Wickramasinghe, Ulapane, Sloane, & Gehlot, 2024). It is a data-intensive process, increasingly supported by advances in Artificial Intelligence (AI), supercomputing, and quantum computing. These technologies are paving the way for next-generation healthcare solutions, including the use of Digital Twins (DTs)—virtual representations of patients that integrate biological and clinical data.

In this chapter, we explore how DTs can be used to uncover complex biological rules hidden within cell-related data, contributing to more precise and personalized care. We focus on a case study involving Triple-Negative Breast Cancer (TNBC), specifically examining how DTs can enhance the planning of immunotherapy treatments involving PD-1/PD-L1 inhibitors.

Immunotherapy for Triple-Negative Breast Cancer (TNBC)

Triple-Negative Breast Cancer (TNBC) is a particularly aggressive form of breast cancer, characterized by the absence of three key receptors: estrogen, progesterone, and HER2. Because of this, TNBC tends to grow faster, spread more easily, and offers fewer treatment options—resulting in a poorer prognosis.

To address this challenge, immune checkpoint inhibitors have emerged as a promising form of therapy. This treatment works by making cancer cells more vulnerable to the body's own immune system. However, this often affects healthy cells too, leading to side effects and complications. This trade-off between effectiveness and safety remains a major concern.

Among these immunotherapies, PD-1/PD-L1 inhibitors have been approved in several countries. Yet, patient responses vary widely, and side effects can be unpredictable. This variability highlights the need to better predict treatment outcomes before therapy begins.

In this chapter, we explore how DTs might help improve the planning of PD-1/PD-L1 treatments for TNBC. By analyzing patterns of statistical congruence in cell-related data, we aim to identify which patients are most likely to benefit through a way of cohort matching—bringing greater precision and personalization to immunotherapy.

DIGITAL TWINS (DTs) IN HEALTHCARE

Digital Twins (DTs) are virtual representations of physical entities, which can include objects, systems, processes, living beings, or even parts of them (Qian et al., 2022). Their primary purpose is to simulate, predict, and optimize various aspects of their real-world counterparts. By continuously communicating with data sources—often in real-time—DTs provide dynamic and accurate models of the physical entities they represent.

Over time, the concept of DTs has evolved and found applications across many industries, such as manufacturing and automation, construction and urban planning, and supply chain and logistics. In healthcare, however, DTs are still a relatively new and emerging concept with significant potential (Katsoulakis et al., 2024; Papachristou et al., 2024).

One of the key goals of using DTs in healthcare is to enhance precision and personalization (Wickramasinghe, Ulapane, Sloane, & Gehlot, 2024) (Cellina et al., 2023; Mokhtari, 2025; Vallée, 2024) and Artificial Intelligence (AI) joins hands in this pursuit (Kaul et al., 2023; Łukaniszyn et al., 2024; Oulefki, Amira, & Foufou, 2025). By integrating data from diverse sources—such as genetic profiles, cell-related data, existing health conditions, lifestyle habits, medical records, real-time sensor inputs, and environmental factors, DTs can create comprehensive and adaptable models of patients and healthcare processes. These models enable simulations that help predict disease progression and tailor treatments to individual patient needs.

For example, DTs can be used to develop personalized treatment plans for chronic illnesses like cancer (Wickramasinghe et al., 2021). By simulating different treatment scenarios based on a patient's current condition, clinicians can identify the most effective options while minimizing risks and side effects. This data-driven approach reduces reliance on trial-and-error methods and supports more evidence-based decision-making, ultimately leading to superior outcomes and improved patient satisfaction.

Research Question

As discussed in the previous subsections, this study examines the potential of DTs to support precision and personalization in healthcare. In particular, we focus on how DTs can improve the planning of PD-1/PD-L1 inhibitor treatments for patients with TNBC.

The central hypothesis behind this work is that complex biological rules—often encoded within our cells—can be uncovered through computational methods and used to tailor treatments to individual patients. By using DTs to capture patient-specific biological patterns, we aim to identify statistically significant cellular markers that may help predict how well a patient will respond to certain immunotherapies.

This leads us to the overarching research question for this chapter: How might the concept of Digital Twins be used to discover complex biological rules to enhance precision and personalization in treatment planning for triple-negative breast cancer?

This question reflects our broader goal: to integrate computational modeling, cell-related data, and clinical decision-making in a way that enables more personalized cancer care. Through this approach, we hope to contribute to the growing conversation around how DTs could reshape the future of personalized care.

COMPLEX BIOLOGICAL RULES

Treatment responses in cancer are often governed by complex, multifactorial, and context-specific biological rules. In the case of TNBC, insights from our discussions with cancer biologists and immunologists have emphasized the role of various immune-related cell types within the tumor microenvironment (TME) as potential indicators of how patients might respond to immunotherapy. The specific cell types relevant to this study are listed in Table 1.

Table 1. Data collection techniques

Symbol	Description
C	Number of Cancer Cells within a unit surface area.
CD3	Number of CD3+ T cells present within the same unit area.
ICOS	Number of activated T cells present within the same unit are (as indicated by ICOS+ expression).
CD8	Number of CD8+ cytotoxic T cells present within the same unit area.
C\CD3	Percentage (%) of CD3+ T cells with respect to the number of cancer cells.
C\ICOS	Percentage (%) of ICOS+ T cells with respect to the number of cancer cells.
C\CD8	Percentage (%) of CD8+ T cells with respect to the number of cancer cells.

Our initial focus is on quantifying the density of cancer cells and three key immune cell types: CD3, ICOS, and CD8. Here's a brief overview:

1. CD3+ cells represent all T cells.
2. ICOS+ cells are a subset of CD3+ T cells that are activated.
3. CD8+ cells are cytotoxic T cells, which play a direct role in attacking cancer cells.

In collaboration with a leading Australian cancer research institute, we use spatial transcriptomics to measure the presence of these immune cells within the TME. At this stage, we concentrate on cell density per unit area. In future phases, we plan to incorporate spatial relationships—such as how close different cell types are to one another—which could provide deeper insights into treatment outcomes.

To illustrate how biological rules might influence treatment responses, we present synthetically generated data in Table 2. This table shows hypothetical rules based on varying cell densities and their associated outcomes under different PD-1/PD-L1 inhibitor treatments. Table 2 includes three example patients (Patient 1, 2, and 3), each represented by a distinct color code and characterized by unique cellular compositions. Based on these conditions, each patient is presumed to respond best to a different immunotherapy regimen.

However, such detailed biological insights are rarely available at the start of treatment. As a result, patients are often assigned therapies without knowing how effective they will be. Our proposed approach addresses this gap by modeling each patient's treatment journey—including therapies received and outcomes observed—as a Digital Twin (DT). As data of more patients is collected over time, patterns may emerge that reveal the biological rules influencing treatment responses.

This method allows for retrospective analysis to guide future decision-making. By identifying statistically consistent patterns between past and new patients, clinicians can better match therapies to individuals, improving the precision and personalization of treatment plans. Ultimately, this approach aims to enhance clinical outcomes—such as overall survival, as shown in Table 2—by using DTs to uncover and apply complex biological rules in real-world healthcare settings.

PROPOSED SOLUTION

The proposed solution features a web-based DT interface, illustrated in Figure 1, designed to support clinical decision-making. This tool is intended for use by clinicians—potentially alongside patients—to enable collaborative and informed treatment planning.

At the top-right corner of the interface is a 'Load Patient' button, which starts the process. When a patient is present, the clinician can click this button to scan a unique patient identifier, such as a QR code. This triggers the secure retrieval of the patient's biomarkers and other relevant clinical data from a dedicated server, which is then displayed on the DT interface.

To protect patient privacy and ensure the accuracy of data, a confirmation popup appears first. It shows limited personal identifiers (e.g., name and date of birth) for verification. These details are only visible in the popup and not on the main screen, maintaining confidentiality. Once verified by both the clinician and the patient, the popup closes, and the patient's data is shown on the DT dashboard.

As seen in Figure 1, the dashboard organizes patient data into two main sections:

1. Demographic Information: Includes age (with a $\pm$5-year range for cohort matching), sex (Male/Female), and ethnicity (e.g., African, East Asian, South Asian, Southeast Asian, MENA, European, Hispanic or Latino, Indigenous Peoples, Pacific Islander, etc.). These attributes help identify a statistically similar group of past patients to guide personalized treatment recommendations.
2. Disease-Specific Details: Tailored for TNBC, this section includes cancer stage (e.g., Stage IA, IB, IIA, etc.) and key biomarkers (listed in Table 1) relevant to PD-1/PD-L1 inhibitor therapies. These biomarkers are treated as quantitative variables, and a $\pm$10% margin is used to match patients with similar profiles.

Clinicians can then choose from available immunotherapy options, such as PD-1 or PD-L1 inhibitors. By clicking the 'Match Cohort' button, the system retrieves outcome data from the matched cohort and displays it in the 'Outcome Measures' panel.

This panel provides key metrics including:

1. Objective Response Rate (ORR)
2. Progression-Free Survival (PFS) range (in months)
3. Overall Survival (OS) range (in months)
4. Likely side effects, such as fatigue, nausea, and immune-related adverse events (AEs)

These insights help clinicians make evidence-based decisions tailored to each patient, enhancing both precision and personalization. The interface also supports shared decision-making, empowering patients through transparency and collaboration.

Importantly, the DT interface acts as a dynamic digital model of the patient, continuously updated with biological, treatment and outcome data. Over time, this enables the identification of statistically significant patterns that can inform future treatment strategies. As shown in Figure 2, we also plan to introduce advanced visualizations that display the distribution of past patients and highlight the most statistically similar cohort for a presenting current patient—further reinforcing the personalized nature of the approach.

Figure 1. Proposed web-based DT dashboard for precise and personalized immunotherapy treatment planning for TNBC

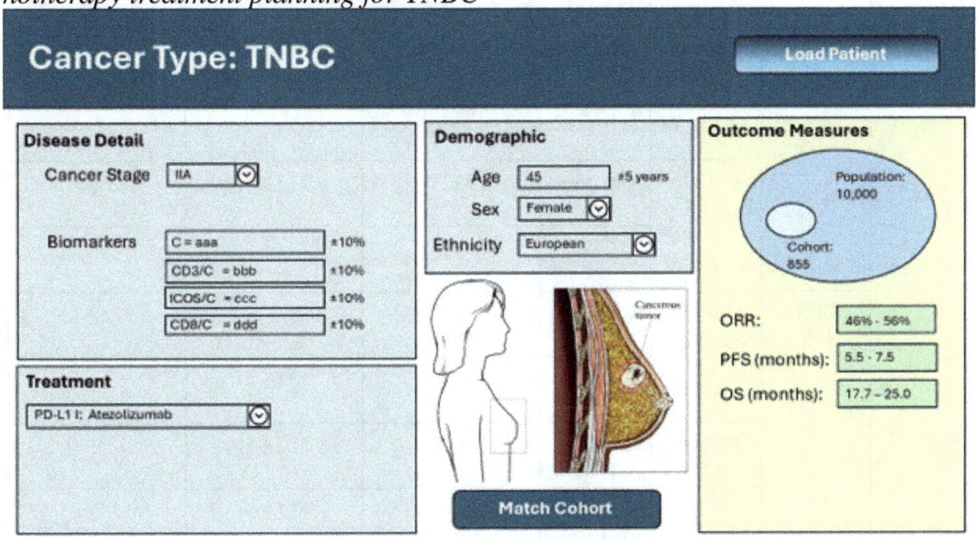

Figure 2. Population of past patients alongside a discovered DT (i.e., the most statistically congruent cohort to a presenting current patient) depicting personalization of outcome measures

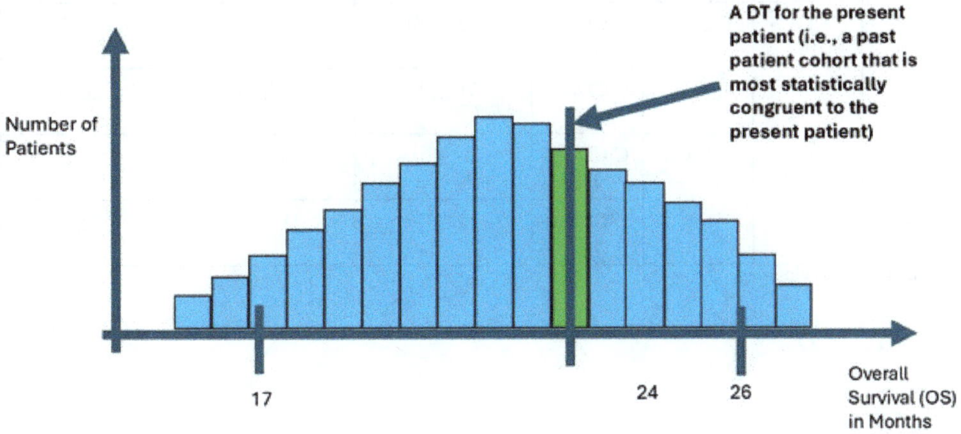

Table 2. Indicative implicit rules (synthetically generated) pertaining to efficacy of various immunotherapies (Legend: Patient 1; Patient 2; Patient 3)

Implicit rule on Cancer Cell Count	Implicit rule on CD3+ T Cell Count	Implicit rule on ICOS+ T Cell Count	Implicit rule on CD8+ T Cell Count	Immunotherapy A (Overall Survival (OS)/ months)	Immunotherapy B (Overall Survival (OS)/ months)	Immunotherapy C (Overall Survival (OS)/ months)
C ≤ 1,000	C\CD3 ≤ 10%	C\ICOS ≤ 5%	C\CD8 ≤ 1%	17 (± 5)	21 (± 5)	13 (± 5)
			C\CD8 > 1%	16 (± 5)	5 (± 5)	8 (± 5)
		C\ICOS > 5%	C\CD8 ≤ 1%	23 (± 5)	11 (± 5)	7 (± 5)
			C\CD8 > 1%	20 (± 5)	24 (± 5)	14 (± 5)
	C\CD3 > 10%	C\ICOS ≤ 5%	C\CD8 ≤ 1%	7 (± 5)	5 (± 5)	22 (± 5)
			C\CD8 > 1%	22 (± 5)	13 (± 5)	6 (± 5)
		C\ICOS > 5%	C\CD8 ≤ 1%	8 (± 5)	31 (± 5)	7 (± 5)
			C\CD8 > 1%	20 (± 5)	11 (± 5)	17 (± 5)
1,000 < C ≤ 10,000	C\CD3 ≤ 10%	C\ICOS ≤ 5%	C\CD8 ≤ 1%	23 (± 5)	28 (± 5)	19 (± 5)
			C\CD8 > 1%	16 (± 5)	30 (± 5)	7 (± 5)
		C\ICOS > 5%	C\CD8 ≤ 1%	6 (± 5)	7 (± 5)	10 (± 5)
			C\CD8 > 1%	14(± 5)	27 (± 5)	11 (± 5)
	C\CD3 > 10%	C\ICOS ≤ 5%	C\CD8 ≤ 1%	12 (± 5)	7 (± 5)	12 (± 5)
			C\CD8 > 1%	21 (± 5)	23 (± 5)	5 (± 5)
		C\ICOS > 5%	C\CD8 ≤ 1%	8 (± 5)	20 (± 5)	5 (± 5)
			C\CD8 > 1%	23 (± 5)	31 (± 5)	16 (± 5)

continued on following page

Table 2. Continued

Implicit rule on Cancer Cell Count	Implicit rule on CD3+ T Cell Count	Implicit rule on ICOS+ T Cell Count	Implicit rule on CD8+ T Cell Count	Immunotherapy A (Overall Survival (OS)/ months)	Immunotherapy B (Overall Survival (OS)/ months)	Immunotherapy C (Overall Survival (OS)/ months)
10,000 > C	C\CD3 ≤ 10%	C\ICOS ≤ 5%	C\CD8 ≤ 1%	10 (± 5)	19 (± 5)	24 (± 5)
			C\CD8 > 1%	14 (± 5)	36 (± 5)	11 (± 5)
		C\ICOS > 5%	C\CD8 ≤ 1%	14 (± 5)	18 (± 5)	10 (± 5)
			C\CD8 > 1%	10 (± 5)	8 (± 5)	18 (± 5)
	C\CD3 > 10%	C\ICOS ≤ 5%	C\CD8 ≤ 1%	5 (± 5)	35 (± 5)	15 (± 5)
			C\CD8 > 1%	5 (± 5)	17 (± 5)	9 (± 5)
		C\ICOS > 5%	C\CD8 ≤ 1%	17 (± 5)	36 (± 5)	5 (± 5)
			C\CD8 > 1%	14(± 5)	6 (± 5)	24 (± 5)

CONCLUSION

This chapter introduced a novel approach to advancing precision and personalization in cancer treatment planning using DTs. Focusing on TNBC and PD-1/ PD-L1 inhibitor therapies, we explored how DTs can help uncover complex biological rules embedded within the tumor microenvironment. By combining spatial transcriptomics with computational modeling, we demonstrated how immune cell densities—particularly CD3+, ICOS+, and CD8+ T cells—can serve as predictive markers for treatment response.

These insights were integrated into a web-based DT interface developed to support clinical decision-making. The interface allows clinicians to match new patients with statistically similar cohorts from historical data, helping guide treatment choices based on real-world outcomes. It also promotes transparency and shared decision-making, empowering patients with personalized insights grounded in evidence.

As more patient data is collected and modeled over time, the DT system is expected to evolve into a powerful tool for discovering and validating biological rules that may influence treatment effectiveness. This continuous learning process not only improves the accuracy of recommendations but also contributes to the broader goal of data-driven, personalized medicine.

In summary, this chapter highlighted the potential of DTs to bridge the gap between complex biological data and actionable clinical insights. By revealing hidden patterns in patient-specific information, DTs offer a promising path toward more effective, personalized, and equitable cancer care, with implications having broader outreach.

REFERENCES

Cellina, M., Cè, M., Alì, M., Irmici, G., Ibba, S., Caloro, E., Fazzini, D., Oliva, G., & Papa, S. (2023). Digital twins: The new frontier for personalized medicine? *Applied Sciences (Basel, Switzerland)*, *13*(13), 7940. DOI: 10.3390/app13137940

Goetz, L. H., & Schork, N. J. (2018). Personalized medicine: Motivation, challenges, and progress. *Fertility and Sterility*, *109*(6), 952–963. DOI: 10.1016/j.fertnstert.2018.05.006 PMID: 29935653

Katsoulakis, E., Wang, Q., & Wu, H.. (2024). Digital twins for health: A scoping review. *npj. Digital Medicine*, *7*, 77. PMID: 38519626

Kaul, R., Ossai, C., Forkan, A. R. M., Jayaraman, P. P., Zelcer, J., Vaughan, S., & Wickramasinghe, N. (2023). The role of AI for developing digital twins in healthcare: The case of cancer care. *Wiley Interdisciplinary Reviews. Data Mining and Knowledge Discovery*, *13*(1), e1480. DOI: 10.1002/widm.1480

Łukaniszyn, M., Majka, Ł., Grochowicz, B., Mikołajewski, D., & Kawala-Sterniuk, A. (2024). Digital twins generated by artificial intelligence in personalized healthcare. *Applied Sciences (Basel, Switzerland)*, *14*(20), 9404. DOI: 10.3390/app14209404

Mokhtari, M. (2025). *Human digital twins in personalized healthcare: An overview and future perspectives*. Unpublished manuscript.

Oulefki, A., Amira, A., & Foufou, S. (2025). Digital twins and AI transforming healthcare systems through innovation and data-driven decision making. *Health and Technology*, *15*(2), 299–321. DOI: 10.1007/s12553-025-00947-x

Papachristou, K., Katsakiori, P. F., Papadimitroulas, P., Strigari, L., & Kagadis, G. C. (2024). Digital twins' advancements and applications in healthcare, towards precision medicine. *Journal of Personalized Medicine*, *14*(11), 1101. DOI: 10.3390/jpm14111101 PMID: 39590593

Qian, C., Liu, X., Ripley, C., Qian, M., Liang, F., & Yu, W. (2022). Digital twin—Cyber replica of physical things: Architecture, applications and future research directions. *Future Internet*, *14*(2), 64. DOI: 10.3390/fi14020064

Vallée, A. (2024). Envisioning the future of personalized medicine: Role and realities of digital twins. *Journal of Medical Internet Research*, *26*, e50204. DOI: 10.2196/50204 PMID: 38739913

Wang, R. C., & Wang, Z. (2023). Precision medicine: Disease subtyping and tailored treatment. *Cancers (Basel), 15*(15), 3837. DOI: 10.3390/cancers15153837 PMID: 37568653

Wickramasinghe, N., Jayaraman, P. P., Forkan, A. R. M., Ulapane, N., Kaul, R., Vaughan, S., & Zelcer, J. (2021). A vision for leveraging the concept of digital twins to support the provision of personalized cancer care. *IEEE Internet Computing, 26*(5), 17–24. DOI: 10.1109/MIC.2021.3065381

Wickramasinghe, N., Ulapane, N., Sloane, E. B., & Gehlot, V. (2024). Digital twins for more precise and personalized treatment. In *MEDINFO 2023—The Future Is Accessible* (pp. 229–233). IOS Press. DOI: 10.3233/SHTI230961

Chapter 7
Generative AI Opportunities in Healthcare

ABSTRACT

The healthcare sector is undergoing profound changes, driven by a convergence of demographic, environmental, social, geopolitical and economic pressures. This paper explores the next-generation challenges in healthcare, including population aging, rising chronic diseases, healthcare workforce shortages, and growing disparities in access and quality of care. In response, several viable strategies—such as public health education, disease prevention, decentralized care models, and chronic disease management—are examined for their potential to mitigate these challenges. The chapter then introduces Generative Artificial Intelligence (Generative AI) as a transformative technology capable of augmenting these strategies. By generating personalized, multilingual, and context-aware content, generative AI can enhance public education, support disease prevention, improve patient self-care, and enable virtual consultations through intelligent avatars.

INTRODUCTION

Modern healthcare has witnessed remarkable advancements over the past century, from the eradication of several deadly diseases to the development of cutting-edge diagnostic and therapeutic technologies. However, as the world evolves, so too do the challenges that face the healthcare systems. These emerging and interconnected challenges (Charlesworth et al., 2015) are reshaping healthcare worldwide and demanding innovative, systemic responses. Some of the modern-day challenges faced by healthcare systems are the following.

DOI: 10.4018/979-8-3373-1877-6.ch007

Demographic Shifts and Population Pressures

One of the most pressing challenges is the rising population worldwide (Sadigov, 2022), which places immense pressure on healthcare infrastructure, resources, and personnel. This is compounded by the aging population, particularly in developed nations, where a growing proportion of citizens are over the age of 65. Older adults typically require more frequent and complex medical care, contributing to increased demand for chronic disease management, long-term care, and geriatric services.

Migration and Cultural Barriers

Simultaneously, rising immigration (Giuntella et al., 2016)—driven by economic opportunity, conflict, and climate change—introduces new complexities. Healthcare systems must now cater to diverse populations with varying languages, cultural norms, and health beliefs. This can lead to communication barriers, misdiagnoses, and inequitable access to care, especially for undocumented or marginalized groups.

Environmental and Geopolitical Stressors

The health of populations is increasingly affected by environmental pollution (Apergis et al., 2020), which contributes to respiratory diseases, cancers, and developmental disorders. In parallel, geopolitical conflicts displace millions, disrupt healthcare infrastructure, and create humanitarian crises where access to even basic medical care becomes a challenge. These stressors also strain the global supply chain for medical goods and pharmaceuticals.

Socioeconomic Inequities

Numerous socioeconomic challenges further exacerbate health disparities (Braveman, 2023). Shortages of quality food and affordable housing, coupled with the rising cost of living and limited access to well-paid jobs, contribute to poor health outcomes. These conditions often lead to malnutrition, chronic stress, and limited access to preventive care, especially in underserved communities.

Social Isolation and Lifestyle Changes

Modern lifestyles have also led to a decline in human connection, with many individuals experiencing family fragmentation and social isolation (Patulny & Bower, 2022). This, combined with addictions (to substances, technology, or dopamine-driven behaviors), contributes to a more sedentary lifestyle and a growing mental health crisis. Depression, anxiety, and loneliness are now recognized as major public health concerns, with significant implications for physical health.

Rise of Chronic Diseases

Numerous lifestyle and environmental factors have contributed to a surge in chronic diseases such as diabetes, cardiovascular disease, and obesity (Balwan & Kour, 2021). Unlike acute illnesses, chronic conditions require long-term management, placing sustained pressure on healthcare systems and increasing the need for personalized, continuous care.

Infrastructure and Workforce Strain

As the demand for healthcare rises, healthcare infrastructure and workforce are struggling to keep pace (Michaeli et al., 2024). There is a decline in available hospital beds per capita, especially in urban centers, leading to overcrowding and delayed treatments. At the same time, the sector faces a shortage of healthcare workers, including doctors, nurses, and allied professionals. Those who remain are often overworked, leading to burnout, reduced job satisfaction, and ultimately, a decline in the quality of care delivered.

Escalating Healthcare Costs

All of the above challenges culminate in rising healthcare costs (Cummings, 2022). From expensive treatments and medications to administrative overheads and insurance complexities, the financial burden on both individuals and governments is growing. This makes healthcare increasingly inaccessible for many, particularly in low- and middle-income countries.

Such next-generation challenges are deeply interconnected and form a complex web of demographic, environmental, social, and economic pressures. They not only strain existing healthcare systems but also highlight the urgent need for innovative, scalable, and equitable solutions. As we look to the future, technologies like Generative AI offer promising avenues to reimagine how care is delivered, accessed, and optimized.

SOME OF THE VIABLE APPROACHES TO ADDRESS THE NEXT-GENERATION HEALTH CHALLENGES

To effectively respond to the multifaceted challenges facing modern healthcare systems, a combination of proactive, patient-centered, and technology-enabled approaches is essential. While no single solution can address all issues, several viable approaches have emerged as promising pathways towards a more resilient, equitable, and sustainable healthcare. These include public education, preventive care, decentralized service delivery, and improved chronic disease management.

Public Education and Awareness of Healthy Lifestyles

Empowering individuals with knowledge is one of the most cost-effective and impactful approaches for achieving effective healthcare (Yu et al., 2022). Public education campaigns and educational digital solutions can promote healthy lifestyles—such as balanced nutrition, regular physical activity, mental well-being, and avoidance of harmful substances. These can significantly reduce the incidence of preventable diseases.

For example, national and digital/online campaigns on smoking cessation, diabetes prevention, or mental health awareness could deliver success in shifting public behavior. Schools, workplaces, and community centers can serve as hubs for health education, while digital platforms and social media can amplify reach and engagement. Importantly, culturally tailored messaging is crucial to ensure inclusivity and effectiveness across diverse populations.

Disease Prevention

Preventive healthcare is essential for a sustainable healthcare system (Hagenaars & Klazinga, 2021). By focusing on early detection, screening programs, and other necessary approaches, healthcare providers can intervene before conditions become severe or chronic. This not only improves patient outcomes but also reduces long-term healthcare costs.

For instance, routine screenings for conditions like hypertension, cancer, and diabetes can catch diseases in their early stages, when they are more manageable and less expensive to treat. Moreover, specialized screenings such as genetic screening and risk profiling are emerging tools that can personalize prevention strategies based on individual risk factors.

Hospital in the Home (HITH) Approaches

The traditional model of hospital-based care is increasingly being supplemented by Hospital in the Home (HITH) (Cubitt & Lim, 2025) programs in modern times. HITH systems aim to deliver acute and subacute care in a patient's residence. This model is particularly effective for patients recovering from conditions such as surgery, infections, or requiring palliative care; but this approach can have more applications.

HITH approaches have several benefits. These include the likes of reduced strain on hospital infrastructure, lower risk of hospital-acquired infections, and enhanced patient comfort and satisfaction. It also allows for more personalized care, often supported by remote monitoring technologies, mobile health teams, and telehealth consultations, and for some, this comes in conjunction with having the luxury of being present with family and loved ones. In regions with limited hospital capacity or during crisis situations like pandemics, HITH approaches can be a vital and effective component enabling healthcare delivery.

Effective Chronic Care Delivery

Chronic diseases such as diabetes, heart disease, cancer and so on are among the leading causes of death and disability worldwide (Hacker, 2024). Addressing these conditions requires a long-term, coordinated approach that goes beyond episodic treatment.

Effective chronic care involves the likes of multidisciplinary teams, patient education, self-management support, and continuous monitoring. Effective chronic care models tend to emphasize proactive planning, evidence-based interventions, and strong patient-provider relationships. Digital tools—such as mobile apps for medication adherence, wearable devices for real-time monitoring, and AI-driven decision support—can further enhance chronic care delivery.

Importantly, care must be equitable and accessible, especially for vulnerable populations who may face barriers due to geography, income, or language. Community health workers, nurse-led clinics, tailored digital solutions and integrated care pathways can help bridge these gaps.

Approaches such as those discussed above—ranging from public education to home-based care—hint at the shift necessary from reactive to proactive healthcare to address the next-generation healthcare challenges. These approaches emphasize on prevention, personalization, and decentralization of healthcare, aligning with the evolving needs of modern populations. However, to scale these solutions effectively, healthcare systems must embrace innovation, invest in infrastructure, and foster cross-sector collaboration. In this context, Generative AI emerges as a powerful tool with the potential to enhance and accelerate the required advancements in healthcare systems.

GENERATIVE AI AND ITS POSSIBLE ROLES IN HEALTHCARE

As healthcare systems grapple with increasingly complex challenges as discussed above, Generative Artificial Intelligence (Generative AI) has emerged as a powerful tool with the potential to reshape how care is delivered, accessed, and experienced (Chatterjee et al., 2024). Unlike traditional AI systems that classify or predict based on existing data, generative AI models tend to create new content—text, images, audio, and even video—based on learned patterns. This capability opens up a wide array of applications in healthcare, particularly relating to the likes of education, communication, personalization, and automation.

Discussed below are several key areas where generative AI can play a crucial role in addressing next-generation healthcare challenges.

Public Health Education and Awareness

Generative AI can be used to create engaging, accessible, and up-to-date health education content for the public. This includes solutions like the following:

1. Infographics, videos, and articles explaining healthy lifestyle choices based on curated input content.
2. Interactive chatbots that answer common health questions.
3. Social media content tailored to different demographics to promote wellness campaigns.

By automating content creation, public health agencies can scale their outreach efforts and keep producing consistent and evidence-based content.

Multilingual and Culturally Sensitive Education Tools

With rising immigration and increasingly diverse populations, language and cultural barriers can hinder effective healthcare delivery and communication. Generative AI can be used through several means to address the associated challenges. Some examples include the following:

1. Translate health-related content into multiple languages while preserving cultural nuances.

2. Generate region-specific examples and analogies to improve comprehension.
3. Support voice-based calling agents or assistants that speak in local dialects or accents.

Such solutions ensure that health education is inclusive and equitable, reaching communities that are often underserved.

Education on Disease Prevention

Generative AI can produce personalized disease prevention strategies through means such as the following.

1. Tailored risk assessments and lifestyle recommendations based on user input.
2. Simulated scenarios showing the long-term impact of preventive behaviors (e.g., quitting smoking).
3. Gamified learning modules that teach users about the likes of nutrition and exercise.

Such solutions can be deployed through the likes of mobile apps, kiosks in clinics, or community health programs.

Disease Management and Self-Care Advice

For patients managing chronic conditions, generative AI can help by producing content such as the following:

1. Daily care instructions and reminders on treatment plans.
2. Conversational agents that answer questions about the likes of symptoms, medications, or diet.
3. Motivational messages to encourage adherence and lifestyle changes.

Such solutions empower patients to take a more active role in their health, reducing dependency on in-person visits.

Hospital Entry and Orientation Tools

Navigating a hospital can be overwhelming, especially for first-time patients. Generative AI can help in this context through means such as the following:

1. Create virtual guides or avatars that explain hospital layouts, departments, and procedures.
2. Offer interactive FAQs for check-in, insurance, and consent processes.
3. Provide real-time translation for non-native speakers during admission.

Such solutions can improve patient experience and reduce the administrative burden on staff.

Hospital Discharge and Post-Discharge Education

Discharge is a critical point where miscommunication can lead to readmissions and entailing complications. Generative AI can help in this context through solutions such those listed below:

1. Generate personalized discharge summaries in plain language.
2. Provide step-by-step recovery instructions, including medication schedules and warning signs.
3. Offer follow-up reminders and virtual check-ins to ensure continuity of care.

Such solutions can be integrated into patient portals or sent via SMS/email.

Nurse and Doctor Avatars

Generative AI can enable realistic, conversational avatars that simulate healthcare professionals. Such avatars can be delivered through smartphone apps, websites, or even through robots. These avatars can help through means such as the following:

1. Conduct virtual consultations for minor or self-manageable conditions.
2. Provide triage support to determine if in-person care is needed.
3. Issue prescriptions or referrals under supervision or integration with licensed providers.

Such solutions can expand access to care, especially in rural or underserved areas.

Therapy and Rehabilitation Education

Generative AI can enhance therapy adherence by enabling the likes of those listed below:

1. Custom physiotherapy videos or text of vocal messages based on patient needs and progress.
2. Rehab programs such as cardiac rehab with interactive feedback and motivational coaching.
3. Mental health support tools, including guided meditation, exercises, and journaling prompts.

Such solutions can be accessed at home by the public, reducing the need for frequent clinic visits.

Additional Emerging Applications

There can be many other applications. Some examples are:

1. Medical training simulations for students and professionals.
2. Conversation training, including the likes of empathy training for healthcare professionals.
3. Clinical documentation assistants that generate notes from voice recordings.
4. Synthetic data generation for research and model training without compromising patient privacy.

Generative AI thus holds immense promise in augmenting human capabilities, bridging communication gaps, and offering personalized care. Its ability to generate content dynamically and contextually makes it a powerful ally in tackling the systemic challenges of modern healthcare. As we explore its potential further, it becomes clear that generative AI is not just a technological innovation, but it is a strategic enabler for a more inclusive, efficient, and patient-centered healthcare of the future.

STRENGTHS, WEAKNESSES, OPPORTUNITIES AND THREATS (SWOT) OF GENERATIVE AI IN HEALTHCARE

As generative AI continues to evolve, its integration into healthcare presents a mix of potential and challenges. A SWOT analysis provides a structured way to evaluate its current and future role in the sector.

Strengths

1. Scalability and Efficiency: Generative AI can produce vast amounts of content—educational materials, clinical summaries, patient instructions—at scale and with minimal human input, significantly reducing administrative burden.
2. Personalization: AI-generated content can be tailored to individual patient profiles, including language, literacy level, cultural context, and medical history, and so on, enhancing patient engagement and comprehension.
3. 24/7 Availability: Virtual assistants and avatars powered by generative AI can provide continuous support (for as long as internet is available in most cases), doing the likes of answering questions, guiding self-care, and offering reassurance outside of clinical hours.
4. Multilingual and Inclusive Communication: Generative AI can bridge language and literacy gaps, making healthcare more accessible to diverse populations through real-time translation and culturally sensitive messaging.
5. Support for Healthcare Professionals: AI can assist clinicians through the likes of generating documentation, summarizing patient histories, and drafting discharge notes, allowing them to focus more on patient care.

Weaknesses

1. Accuracy and Reliability: Generative AI models can sometimes produce plausible but incorrect or misleading information, which can be dangerous in a healthcare context if not properly validated.
2. Lack of Clinical Judgment: AI lacks the nuanced understanding and ethical reasoning of human clinicians, making it limited in complex decision-making without human oversight.
3. Data Privacy and Security Risks: The use of patient data to train or operate generative models raises concerns about data protection, consent, and raises the need for compliance with regulations.
4. Bias and Inequity: If trained on biased or non-representative data, generative AI can perpetuate or even amplify existing health disparities, particularly for marginalized groups.
5. Overdependence on Technology: Excessive reliance on AI tools may lead to deskilling healthcare workers or reduced human interaction, which is vital for patient trust and empathy.

Opportunities

1. Expanding Access to Care: Generative AI can extend healthcare services to remote, rural, and underserved populations, reducing geographic and economic barriers.
2. Enhancing Preventive Care: AI-generated educational tools can promote early intervention and lifestyle changes, reducing the burden of chronic diseases.
3. Training and Simulation: AI can create realistic medical training scenarios, helping students and professionals practice communication, diagnosis, and emergency response.
4. Mental Health Support: Generative AI can power chatbots and virtual therapists that provide the likes of cognitive behavioral therapy, mindfulness exercises, and emotional support.
5. Innovation in Research and Development: Interactions through generative AI can assist the likes of drug discovery as per public needs, clinical trial design, choice of right candidates for clinical trials, and synthetic data generation, accelerating innovation while protecting patient privacy.

Threats

1. Regulatory and Legal Uncertainty: The rapid evolution of generative AI outpaces current regulatory frameworks, creating ambiguity around liability, accountability, and ethical use.
2. Misinformation and Misuse: Malicious actors could use generative AI to create fake medical advice, fraudulent prescriptions, or deepfake consultations, undermining public trust.
3. Resistance to Adoption: Healthcare professionals may be skeptical or resistant to integrating AI into their workflows due to concerns about job displacement, accuracy, or loss of autonomy.
4. Infrastructure and Cost Barriers: Implementing generative AI solutions requires robust digital infrastructure, training, and ongoing maintenance, which may be challenging for low-resource settings.
5. Ethical Dilemmas: Questions around consent, transparency, and human oversight remain unresolved as yet, especially when AI is involved in sensitive or life-altering decisions.

REFERENCES

Apergis, N., Bhattacharya, M., & Hadhri, W. (2020). Health care expenditure and environmental pollution: A cross-country comparison across different income groups. *Environmental Science and Pollution Research International, 27*(8), 8142–8156. DOI: 10.1007/s11356-019-07457-0 PMID: 31897990

Balwan, W. K., & Kour, S. (2021). Lifestyle Diseases: The Link between Modern Lifestyle and threat to public health. *Saudi J Med Pharm Sci, 7*(4), 179–184. DOI: 10.36348/sjmps.2021.v07i04.003

Braveman, P. (2023). *The social determinants of health and health disparities.* Oxford University Press. DOI: 10.1093/oso/9780190624118.001.0001

Charlesworth, K., Jamieson, M., Butler, C. D., & Davey, R. (2015). The future healthcare? *Australian Health Review, 39*(4), 444–447. DOI: 10.1071/AH14243 PMID: 25818333

Chatterjee, S., Fruhling, A., Kotiadis, K., & Gartner, D. (2024). Towards new frontiers of healthcare systems research using artificial intelligence and generative AI. *Health Systems (Basingstoke, England), 13*(4), 263–273. DOI: 10.1080/20476965.2024.2402128 PMID: 39584173

Cubitt, M., & Lim, S. (2025). A future for the hospital-in-the-home (HITH) deteriorating patient: Shifting the paradigm. *The Medical Journal of Australia, 222*(4), 168–171. DOI: 10.5694/mja2.52588 PMID: 39844507

Cummings, B. (2022). Rising Healthcare Costs Are a Rising Concern. *Journal of Financial Planning, 35*(2), 19.

Giuntella, O., Nicodemo, C., & Vargas Silva, C. (2016). The impact of immigration on health and health care: Evidence from the United Kingdom. *Refugees and economic migrants: Facts, policies and challenges*, 99-114.

Hacker, K. (2024). The burden of chronic disease. *Mayo Clinic Proceedings. Innovations, Quality & Outcomes, 8*(1), 112–119. DOI: 10.1016/j.mayocpiqo.2023.08.005 PMID: 38304166

Hagenaars, L. L., & Klazinga, N. S. (2021). Prevention as a strategy for fiscally sustainable healthcare. *Reducing paper and sugar for fiscally sustainable healthcare systems*, 95.

Michaeli, D. T., Michaeli, J. C., Albers, S., & Michaeli, T. (2024). The health-care workforce shortage of nurses and physicians: Practice, theory, evidence, and ways forward. *Policy, Politics & Nursing Practice*, *25*(4), 216–227. DOI: 10.1177/15271544241286083 PMID: 39396540

Patulny, R., & Bower, M. (2022). Beware the "loneliness gap"? Examining emerging inequalities and long-term risks of loneliness and isolation emerging from COVID-19. *The Australian Journal of Social Issues*, *57*(3), 562–583. DOI: 10.1002/ajs4.223 PMID: 35942301

Sadigov, R. (2022). Rapid growth of the world population and its socioeconomic results. *The Scientific World Journal*, *2022*(1), 8110229. DOI: 10.1155/2022/8110229 PMID: 35370481

Yu, D. S. F., Li, P. W. C., Li, S. X., Smith, R. D., Yue, S. C. S., & Yan, B. P. (2022). Effectiveness and cost-effectiveness of an empowerment-based self-care education program on health outcomes among patients with heart failure: A randomized clinical trial. *JAMA Network Open*, *5*(4), e225982. DOI: 10.1001/jamanetworkopen.2022.5982 PMID: 35380643

Chapter 8
Agentic AI Opportunities

ABSTRACT

The healthcare sector is undergoing significant changes, driven by intersecting demographic, environmental, social, geopolitical, and economic forces. This chapter examines emerging challenges such as aging populations, increasing prevalence of chronic diseases, shortages in the healthcare workforce, and widening gaps in access and quality of care. In response, we evaluate strategic approaches such as public health education, preventive care, decentralized service models, and chronic disease management as potential solutions.

INTROUDCTION

Agentic Artificial Intelligence (Agentic AI) describes AI systems capable of autonomous, goal-driven behavior, with the ability to make decisions, adapt to dynamic environments, and act proactively to fulfill defined objectives. Unlike conventional AI, which typically operates within narrower scenarios, agentic AI is designed to initiate actions, collaborate with users, and evolve through continuous learning.

This chapter presents Agentic AI as a tool for healthcare, enabling intelligent, context-aware automation across various domains. These systems can support tasks such as public education, enhancing disease prevention, empowering patient self-management, assisting clinical decision-making, and delivering virtual care through intelligent agents. Applications include the likes of adaptive therapy, proactive health coaching, autonomous triage, and workflow optimization.

We present a common framework that can be adopted to design Agentic AI tools for healthcare and discuss various scenario examples as to how Agentic AI can bring about a transformation in healthcare. While offering scalability, adaptability, and continuous learning, Agentic AI also introduces challenges around safety, accountability, data governance, and ethical oversight—we summarize some of these issues in a SWOT (Strengths, Weaknesses, Opportunities and Threats) analysis. The

DOI: 10.4018/979-8-3373-1877-6.ch008

chapter concludes by advocating for responsible integration of agentic AI to foster a more inclusive, efficient, and patient-centered healthcare future.

EMERGING CHALLENGES IN HEALTHCARE

Over the past century, healthcare has made immense progress—from eliminating deadly diseases to pioneering advanced diagnostic and therapeutic tools. Yet, as conditions evolve worldwide, healthcare systems have been facing a new wave of complex, interrelated challenges (Charlesworth et al., 2015) that demand innovative and systemic solutions.

Demographic Shifts and Population Growth

A major concern is the rise in population worldwide (Sadigov, 2022), which intensifies pressure on healthcare infrastructure, resources, and personnel. This is further complicated by aging demographics, especially in developed countries, where an increasing segment of the population is over 65. Older adults typically require more frequent and specialized care, increasing demand for chronic disease management, aged care, and geriatric services.

Migration and Cultural Diversity

Increasing migration (Giuntella et. al., 2016), driven by economic, environmental, and geopolitical factors, introduces new complexities. Healthcare systems must now serve culturally diverse populations with varied languages, health beliefs, and expectations. These differences can lead to communication gaps, misdiagnoses, and unequal access to care, particularly for underrepresented or marginalized groups.

Environmental and Geopolitical Challenges

Environmental pollution and degradation (Apergis et al., 2020) contributes to a rise in various health conditions such as respiratory illnesses, cancers, and developmental disorders. Simultaneously, geopolitical instability displaces populations, disrupts healthcare infrastructure, and creates humanitarian crises where even basic medical services are scarce. These factors also impact global supply chains for medical equipment and pharmaceuticals.

Socioeconomic Disparities

Widening socioeconomic gaps (Braveman, 2023) exacerbate health inequalities. Limited access to nutritious food, affordable housing, and stable employment leads to poor health outcomes. These conditions foster malnutrition, chronic stress, and reduced access to preventive and other healthcare services, particularly in low-income and underserved communities.

Social Isolation and Lifestyle Issues

Modern lifestyles have led to increased social isolation and family fragmentation (Patulny & Bower, 2022). Combined with rising addictions, whether to substances, technology, or dopamine-driven behaviors—these have contributed to sedentary habits and a growing mental health crisis. Depression, anxiety, and loneliness are now major public health issues with significant physical health implications.

Chronic Disease Burden

Environmental and lifestyle factors have driven a surge in chronic illnesses such as diabetes, cardiovascular disease, and obesity (Balwan & Kour, 2021). These conditions require long-term, personalized care, placing sustained pressure on healthcare systems and increasing the need for continuous management strategies.

Infrastructure and Workforce Limitations

Healthcare infrastructure and workforce capacity are struggling to meet the rising modern demands (Michaeli et al., 2024). Urban centers face declining hospital bed availability, leading to overcrowding and treatment delays. Simultaneously, shortages of healthcare professionals—doctors, nurses, and allied staff—result in burnout and reduced care quality.

Rising Healthcare Costs

All such challenges discussed above contribute to escalating healthcare costs (Cummings, 2022). From high treatment and medication expenses to administrative inefficiencies and insurance complexities, the financial burden on individuals, organizations and governments continues to grow, making healthcare less accessible, especially in low- and middle-income regions.

These interconnected challenges form a complex web of demographic, environmental, social, and economic pressures. They strain existing healthcare systems and emphasize the urgent need for scalable, equitable, and innovative solutions. Technologies like Generative AI offer promising pathways to reimagine care delivery, access, and optimization for the future.

VIABLE APPROACHES TO ADDRESS EMERGING HEALTH CHALLENGES

Addressing the complex challenges facing the modern healthcare systems requires a multifaceted approach that is proactive, patient-focused, and technology-driven. While no single solution can resolve all issues, several promising approaches have emerged to make the healthcare systems more resilient, equitable, and sustainable. These include the likes of public health education, preventive care, decentralized service models, and enhanced chronic disease management.

Public Health Education and Awareness of Healthy Lifestyles

Educating individuals is among the most cost-effective and impactful strategies for improving health outcomes (Yu et al., 2022). Public campaigns and digital education tools can promote healthier lifestyles—encouraging balanced diets, regular exercise, mental wellness, and avoidance of harmful habits. These efforts can significantly reduce the prevalence of preventable diseases.

National and online initiatives targeting the likes of smoking cessation, diabetes prevention, and mental health awareness have shown success in shifting public behavior. Schools, workplaces, and community hubs can serve as centers for health education, while digital platforms and social media extend reach and engagement. Tailoring messages to cultural contexts is essential to ensure inclusivity and effectiveness across diverse populations.

Preventive Healthcare

Preventive care is a cornerstone of sustainable healthcare systems (Hagenaars & Klazinga, 2021). Early detection through screening programs allows for timely intervention before conditions become severe or chronic, improving outcomes and reducing long-term costs.

Routine screenings for hypertension, cancer, and diabetes help identify diseases in their early stages. Emerging tools like genetic profiling and risk-based assessments enable personalized prevention strategies, aligning care with individual health risks.

Hospital in the Home (HITH) Models

Traditional hospital-based care is increasingly complemented by Hospital in the Home (HITH) programs (Cubitt & Lim, 2025), which deliver acute and subacute care directly in patients' homes. These models are effective for scenarios like post-surgical recovery, infection management, and palliative care, among other applications.

HITH offers multiple benefits: reduced hospital congestion, lower risk of hospital-acquired infections, and improved patient comfort. Supported by remote monitoring, mobile health teams, and telehealth consultations, HITH enables personalized care while allowing patients to remain close to family. In areas with limited hospital capacity or during crises like pandemics, HITH becomes a critical tool for maintaining care delivery.

Enhanced Chronic Disease Management

Chronic conditions such as diabetes, cardiovascular disease, and cancer are leading causes of mortality and morbidity worldwide (Hacker, 2024). Managing these diseases requires coordinated, long-term care beyond episodic treatment.

Effective chronic care involves multidisciplinary teams, patient education, self-management support, and continuous monitoring. Successful models emphasize proactive planning, evidence-based practices, and strong patient-provider relationships. Digital tools—like medication tracking apps, wearable health monitors, and AI-powered decision support—can further enhance care delivery.

Equity in chronic care is vital. Vulnerable populations often face barriers due to geography, income, or language. Community health workers, nurse-led clinics, culturally adapted digital tools, and integrated care pathways can help close these gaps.

Approaches alike those discussed above—from public education to home-based care—reflect a necessary shift from reactive to proactive healthcare. The discussed approaches prioritize prevention, personalization, and decentralization, aligning with the evolving needs of modern populations. To scale these solutions effectively, healthcare systems must embrace innovation, invest in infrastructure, and foster cross-sector collaboration. In this context, Agentic AI stands out as a technology capable of accelerating and enhancing these advancements.

AGENTIC AI AND ITS POSSIBLE APPLICATIONS IN HEALTHCARE

As healthcare systems face increasingly complex challenges—such as aging populations, rising chronic diseases, workforce shortages, and disparities in care—

Agentic Artificial Intelligence (Agentic AI) is emerging as a transformative force in how care is delivered, accessed, and experienced. Unlike traditional AI systems that operate reactively or within narrow constraints, agentic AI systems are designed to act autonomously, pursue goals, adapt to changing contexts, and collaborate with users. These capabilities open up a wide array of applications in healthcare, particularly in areas like education, decision support, personalization, and automation.

Below are several key areas where agentic AI can play a crucial role in addressing next-generation healthcare challenges:

Public Health Education and Awareness

Agentic AI can autonomously deliver timely, personalized health education by:

1. Proactively generating and distributing health content tailored to individual or community needs.
2. Engaging users through interactive agents that answer questions and guide behavior change.
3. Monitoring public health trends and adapting messaging accordingly.

These enable public health agencies to scale outreach and maintain relevance in dynamic environments.

Multilingual and Culturally Sensitive Education Tools

Agentic AI can bridge communication gaps by:

1. Translating and adapting health content across languages and cultural contexts.
2. Initiating conversations in local dialects through voice-based agents.
3. Learning from user interactions to improve cultural sensitivity over time.

These tools ensure inclusive and equitable access to health information.

Education on Disease Prevention

Agentic AI can support preventive care by:

1. Conducting autonomous risk assessments and offering personalized recommendations.
2. Simulating future health scenarios to motivate behavior change.
3. Delivering gamified, adaptive learning experiences based on user engagement.

Such solutions can be embedded in mobile apps, kiosks, or community programs.

Disease Management and Self-Care Advice

For chronic disease management, agentic AI can:

1. Monitor patient data and autonomously adjust care instructions.
2. Engage in ongoing dialogue to answer questions and provide emotional support.
3. Reinforce adherence through motivational coaching and timely reminders.

These empower patients to manage their health more effectively between clinical visits.

Hospital Entry and Orientation Tools

Agentic AI can enhance patient experience by:

1. Acting as virtual guides that navigate hospital layouts and procedures.
2. Proactively assisting with check-in, insurance, and consent processes.
3. Offering real-time translation and support for non-native speakers.

These agents reduce stress and administrative burden for both patients and staff.

Hospital Discharge and Post-Discharge Education

Agentic AI can ensure continuity of care by:

1. Generating and explaining personalized discharge instructions.
2. Monitoring recovery progress and initiating follow-up actions.
3. Alerting clinicians to potential complications based on patient feedback.

These help in reducing readmissions and improving post-discharge outcomes.

Nurse and Doctor Agents

Agentic AI can simulate healthcare professionals through:

1. Conducting virtual consultations and triage autonomously.
2. Collaborating with licensed providers to issue the likes of prescriptions or referrals.

3. Learning from interactions to improve diagnostic and communication skills.

Such agents expand access to care, especially in underserved regions.

Therapy and Rehabilitation Education and Adherence

Agentic AI can support therapy adherence by:

1. Delivering personalized rehab programs with real-time feedback.
2. Adjusting exercises based on patient progress and preferences.
3. Providing mental health support through interactive coaching and journaling.

Such tools promote recovery and reduce reliance on in-person sessions.

Additional Emerging Applications

Agentic AI opens doors to many other innovations, such as:

1. Autonomous medical training simulations for students and professionals.
2. Empathy and communication training for healthcare providers.
3. Intelligent clinical documentation agents that learn from voice interactions.
4. Intelligent clinical decision support tools.
5. Intelligent data connectivity and access tools.
6. Synthetic patient modeling for research and ethical AI development.

Agentic AI thus holds potential to augment human capabilities in healthcare, enabling proactive care, decision support and fostering personalization. The ability of Agentic AI to act independently, adapt intelligently, and collaborate meaningfully positions it as a strategic enabler for a more inclusive, efficient, and patient-centered healthcare future.

A SUGGESTED AGENTIC AI FRAMEWORK

As discussed earlier, Agentic AI has potential to help with several healthcare tasks. These include public (and clinical) education, increasing awareness of healthy lifestyles—which is also tied with preventive healthcare and chronic disease management—and of course in many healthcare operational tasks including clinical decision support. In this section we propose a generic framework that can be used to implement AI agents to assist in healthcare. Our framework combines Natural

Language Processing (NLP), Generative AI (GenAI) and Agentic AI. The proposed framework is presented in Figure 1.

Framework Overview

At its core, the framework begins with an end-user-facing frontend, which could be accessed via a desktop computer or smartphone. The end user could be any member of the public, such as a patient, a member of the public seeking health advice, a clinician, or a student. Through a dedicated interface, users can pose questions in plain language.

These queries are processed by an NLP layer, which decodes the prompts into structured data formats. This layer also performs privacy-preserving preprocessing, such as removing personally identifiable information (PII) to ensure compliance with data protection and privacy standards.

The structured query is then passed to the Agentic AI layer, where an Administrator Agent orchestrates the response process. This Administrator Agent identifies the nature of the query and activates the appropriate Expert AI Agents—for example, a cardiology-related question would trigger a cardiologist-mimicking agent.

The Expert Agent then interacts with the Data Layer, which contains curated, validated, and regulated medical knowledge. This data is sourced from trusted repositories (i.e., best available evidence, or agreed upon information) and reviewed by human experts to ensure accuracy and relevance. The Expert Agent scans the data and retrieves the most pertinent information, operating under strict rule-based algorithms to minimize errors.

The retrieved information is sent back to the Administrator Agent, which applies quality assurance checks and guardrails to assess the reliability and appropriateness of the response. If the information passes these checks, it is forwarded to the Generative AI layer, which converts it into human-readable natural language and delivers it to the user.

If the quality checks fail, the query is escalated to an Escalation Agent, which notifies human experts responsible for oversight, retraining, and data curation. With user consent, conversations may be recorded in the Data Layer for future improvements and quality assurance. Users are also invited to rate the responses, and feedback is stored to refine the system. In cases of low ratings or flagged concerns, the Escalation Agent can forward the case to human experts for review.

Figure 1. A suggested agentic AI framework for healthcare

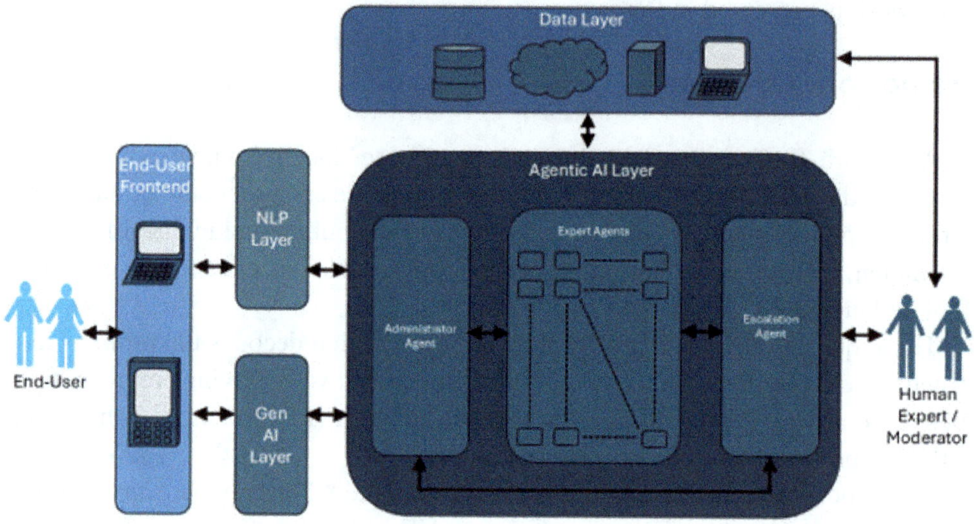

Example Applications

Public Health Education

A member of the public may ask: "What are the early signs of heart disease?"

The frontend captures the query, and the NLP layer decodes it. The Administrator Agent activates the cardiology Expert Agent, which retrieves validated information from the Data Layer. After quality checks, the GenAI layer may present the answer: "Early signs of heart disease may include chest discomfort, shortness of breath, fatigue, and irregular heartbeat. If you experience these symptoms, consult a healthcare provider."

The user may rate the response positively, and the interaction is stored for future refinement.

Patient Self-Management (Diabetes Scenario)

A patient with Type 2 diabetes uses the system and may ask: "What should I eat today to keep my blood sugar stable?"

The system activates a diabetes Expert Agent, which accesses dietary guidelines and personalized data (if available). The response might be: "Consider a balanced meal with whole grains, lean protein, and non-starchy vegetables. Avoid sugary drinks and processed snacks. Monitor your glucose levels before and after meals."

The system might also send reminders, track adherence, quiz the patient for more personalization (for instance, ask for blood sugar levels and the time it was tested and also information about previous meals, drinks and snacks), and also escalate concerns to a care team if needed.

Clinical Decision Support (Cancer Treatment Planning)

An oncologist may ask: "What are the recommended treatment options for stage II colorectal cancer in a 65-year-old female patient?"

The system activates an oncology Expert Agent, which retrieves evidence-based guidelines and recent clinical trial data. After validation, the GenAI layer may respond:

"Standard treatment includes surgical resection followed by adjuvant chemotherapy, typically FOLFOX regimen. Consider patient comorbidities and preferences when planning."

Such responses can be helpful for informed decision-making while maintaining clinical autonomy.

Medical Education (Student Scenario)

A medical student preparing for exams may ask: "What is the difference between systolic and diastolic heart failure?"

The system may respond: "Systolic heart failure involves reduced ejection fraction due to impaired ventricular contraction. Diastolic heart failure involves preserved ejection fraction but impaired ventricular relaxation and filling."

The student may rate the answer and bookmark it for revision.

Merits and Benefits

A system such as the proposed one can have numerous merits and benefits, and some are listed here:

1. Scalability: Can serve millions of users simultaneously.
2. Personalization: Can Tailor responses to individual needs and contexts.
3. Privacy: Built-in mechanisms can perform anonymization and data protection.
4. Efficiency: Can reduce burden on healthcare professionals and systems.
5. Continuous Learning and Improvement: Can improve over time through feedback and expert oversight.

6. Equity: Accessible to diverse populations, including those in remote areas, and can be designed to be multilingual and culturally sensitive.

Limitations and Challenges

1. Data Quality: Requires rigorous curation and validation of medical data.
2. Ethical Oversight: Must ensure transparency, accountability, and fairness.
3. User Trust: Building confidence in AI-generated responses is critical; the proposed system can be made by design, to have fewer black-box elements, and more explainable and rule-based elements.
4. Integration: Needs seamless interoperability with existing healthcare systems.
5. Resource Requirements: High computational and resource investment for development and maintenance.

STRENGTHS, WEAKNESSES, OPPORTUNITIES AND THREATS (SWOT) OF AGENTIC AI IN HEALTHCARE

Like any other technology, the integration of Agentic AI into healthcare presents a mix of potential and challenges. Considering Agentic AI in healthcare more broadly, the SWOT analysis below lists some of the relevant issues facing agentic AI in healthcare.

Strengths

1. Autonomous Decision-Making: Agentic AI systems can act independently to achieve defined healthcare goals, reducing the need for constant human input and enabling proactive care delivery.
2. Context-Aware Personalization: These systems can adapt to individual patient contexts—medical history, preferences, environment—offering highly personalized and responsive support.
3. Continuous Learning and Adaptation: Agentic AI can learn from interactions and outcomes over time, improving its performance and relevance in dynamic healthcare settings.
4. Workflow Optimization: Agentic AI agents can manage routine tasks, triage patients, and coordinate care pathways, freeing up clinicians to focus on complex cases.
5. Scalable Virtual Assistance: Intelligent agents can operate 24/7, providing consistent support across multiple touchpoints—apps, kiosks, wearables—enhancing accessibility and patient engagement.

Weaknesses

1. Complexity of Implementation: Designing and deploying agentic AI systems requires advanced infrastructure, integration with clinical systems, and robust governance frameworks.
2. Limited Clinical Intuition: While agentic AI can make decisions, it may lack the deep clinical judgment and empathy of human professionals, which may limit its role in nuanced scenarios.
3. Data Dependency and Quality Issues: Agentic AI relies heavily on high-quality, real-time data. Incomplete or biased data can impair its decision-making and lead to suboptimal outcomes.
4. Transparency and Explainability Challenges: The autonomous nature of agentic AI can make it difficult to trace or explain decisions, raising concerns in clinical accountability and patient trust.
5. Interoperability Barriers: Integrating agentic AI with diverse healthcare systems and standards can be technically challenging and resource-intensive.

Opportunities

1. Proactive and Preventive Care: Agentic AI can monitor patient data and initiate early interventions, reducing hospitalizations and improving chronic disease management.
2. Remote and Decentralized Healthcare Delivery: Intelligent agents can extend care to remote and underserved areas, supporting virtual consultations, triage, and follow-up care.
3. Personalized Therapy and Rehabilitation: Agentic AI can tailor therapy plans, monitor progress, and adjust interventions in real time, enhancing recovery and adherence.
4. Clinical Decision Support: These systems can assist clinicians by synthesizing data, suggesting treatment options, and flagging potential risks, improving diagnostic accuracy.
5. Health System Resilience: Agentic AI can help manage healthcare operations during crises—pandemics, disasters—by autonomously reallocating resources and coordinating responses.

Threats

1. Ethical and Legal Ambiguities: Autonomous decision-making raises complex questions around liability, consent, and the boundaries of machine authority in healthcare.

2. Security and Privacy Risks: Continuous data collection and decision-making increase exposure to cybersecurity threats and privacy breaches.
3. Trust and Acceptance Issues: Patients and clinicians may be hesitant to rely on autonomous systems, especially in high-stakes or emotionally sensitive situations.
4. Bias and Inequity: If not carefully designed, agentic AI may reinforce systemic biases, particularly in underserved or minority populations.
5. Regulatory Lag: Existing healthcare regulations may not adequately address the capabilities and risks of agentic AI, slowing adoption and innovation.

CONCLUSION

As explored throughout this chapter, Agentic AI offers great potential in reshaping healthcare delivery, education, and accessibility by introducing autonomous, goal-oriented, and context-sensitive systems. However, its effective adoption hinges on thoughtful design, rigorous validation, inclusive nature, and solid yet transparent and fair governance. Given the on going learning capabilities of Agentic Ai this enables new knowledge to be developed and enfolded into current approaches; thereby keeping solutions as relevant and precise as possible. By leveraging its strengths, addressing inherent limitations, capitalizing on emerging opportunities, and proactively managing associated risks, stakeholders can unlock the potential of agentic AI to create a more personalized, efficient, inclusive, and patient-focused healthcare systems for the future.

REFERENCES

Apergis, N., Bhattacharya, M., & Hadhri, W. (2020). Health care expenditure and environmental pollution: A cross-country comparison across different income groups. *Environmental Science and Pollution Research International, 27*(8), 8142–8156. DOI: 10.1007/s11356-019-07457-0 PMID: 31897990

Balwan, W. K., & Kour, S. (2021). Lifestyle Diseases: The Link between Modern Lifestyle and threat to public health. *Saudi J Med Pharm Sci, 7*(4), 179–184. DOI: 10.36348/sjmps.2021.v07i04.003

Braveman, P. (2023). *The social determinants of health and health disparities.* Oxford University Press. DOI: 10.1093/oso/9780190624118.001.0001

Charlesworth, K., Jamieson, M., Butler, C. D., & Davey, R. (2015). The future healthcare? *Australian Health Review, 39*(4), 444–447. DOI: 10.1071/AH14243 PMID: 25818333

Cubitt, M., & Lim, S. (2025). A future for the hospital-in-the-home (HITH) deteriorating patient: Shifting the paradigm. *The Medical Journal of Australia, 222*(4), 168–171. DOI: 10.5694/mja2.52588 PMID: 39844507

Cummings, B. (2022). Rising Healthcare Costs Are a Rising Concern. *Journal of Financial Planning, 35*(2), 19.

Giuntella, O., Nicodemo, C., & Vargas Silva, C. (2016). The impact of immigration on health and health care: Evidence from the United Kingdom. Refugees and economic migrants: Facts, policies and challenges, 99-114.

Hacker, K. (2024). The burden of chronic disease. *Mayo Clinic Proceedings. Innovations, Quality & Outcomes, 8*(1), 112–119. DOI: 10.1016/j.mayocpiqo.2023.08.005 PMID: 38304166

Hagenaars, L. L., & Klazinga, N. S. (2021). Prevention as a strategy for fiscally sustainable healthcare. Reducing paper and sugar for fiscally sustainable healthcare systems, 95.

Michaeli, D. T., Michaeli, J. C., Albers, S., & Michaeli, T. (2024). The healthcare workforce shortage of nurses and physicians: Practice, theory, evidence, and ways forward. *Policy, Politics & Nursing Practice, 25*(4), 216–227. DOI: 10.1177/15271544241286083 PMID: 39396540

Patulny, R., & Bower, M. (2022). Beware the "loneliness gap"? Examining emerging inequalities and long-term risks of loneliness and isolation emerging from COVID-19. *The Australian Journal of Social Issues*, *57*(3), 562–583. DOI: 10.1002/ajs4.223 PMID: 35942301

Sadigov, R. (2022). Rapid growth of the world population and its socioeconomic results. *The Scientific World Journal*, *2022*(1), 8110229. DOI: 10.1155/2022/8110229 PMID: 35370481

Yu, D. S. F., Li, P. W. C., Li, S. X., Smith, R. D., Yue, S. C. S., & Yan, B. P. (2022). Effectiveness and cost-effectiveness of an empowerment-based self-care education program on health outcomes among patients with heart failure: A randomized clinical trial. *JAMA Network Open*, *5*(4), e225982. DOI: 10.1001/jamanetworkopen.2022.5982 PMID: 35380643

Section 3
The Process Perspective

Chapter 9
Strategic Knowledge Management

ABSTRACT

This chapter examines strategic knowledge management (SKM) in healthcare operations, presenting frameworks for aligning knowledge assets with organizational objectives to enhance healthcare delivery and operational excellence. The chapter explores generic KM strategies, industry dynamics through Porter's Five Forces, internal capabilities via Value Chain analysis, root causes of KM challenges using Fishbone analysis, and technology alignment through McFarlan's Strategic Grid which can be considered as foundational concepts of knowledge environments and constructs. A structured approach to KM strategy development is presented including strategic alignment, capability assessment, framework development, and implementation planning. As healthcare evolves toward more technology-enabled models, SKM represents a critical capability for organizations seeking to succeed in progressive complex environments while leveraging collective intelligence to improve outcomes and create sustainable value.

INTRODUCTION

The healthcare sector is currently experiencing substantial transformation in terms of technological advancements, shifting demographics, and evolving and complex patient expectations. The center of this transformation is the strategic management of knowledge, which has emerged as a critical determinant of organizational performance and competitive advantage (Mahdi et al., 2011; Omotayo, 2015). As healthcare systems worldwide manage with the dual challenges of cost

DOI: 10.4018/979-8-3373-1877-6.ch009

containment and quality improvement however, the effective management of organizational knowledge has been given less priority.

Strategic knowledge management (SKM) represents the integration of knowledge management (KM) principles with strategic planning processes to enhance organizational decision-making and value creation (Carayannis et al., 2021). SKM involves the systematic planning, implementation, and evaluation of knowledge processes that align with the organization's strategic objectives and contribute to the delivery of high quality, patient-centered care in the healthcare context (Pflugfelder, 2021). The strategic dimension of KM is crucial as it ensures that knowledge initiatives are not practiced in isolation but are embedded within the broader strategic framework of the healthcare organization.

The relevance of SKM in healthcare is understood by several interconnected trends. The rapid rise of medical information has created a "knowledge explosion" that poses significant challenges for healthcare practitioners in staying current with best practices (Bobruk et al., 2023). Also, the emergence of value-based care models has shifted the focus from volume to value, forcing more sophisticated approaches to knowledge utilization and dissemination (Zanotto et al., 2021). The increasing complexity of healthcare delivery, characterized by multi-disciplinary teams care processes, requires robust knowledge integration mechanisms to ensure care coordination and continuity (Colombani et al., 2022).

Despite the recognized importance of SKM, many healthcare organizations struggle with its implementation. Healthcare organizations worldwide increasingly recognize the importance of KM in improving patient outcomes and operational efficiency. Siemens Healthcare developed a knowledge-based system integrating hospitals, clinics, and pharmacies, allowing for enhanced collaboration and improved healthcare delivery (Stroetmann & Aisenbrey, 2012).

This chapter aims to address this gap by providing a comprehensive framework for developing and implementing a strategic KM approach in healthcare organizations. Based on both theoretical foundations and empirical evidence, the chapter explores various strategic models and analytical tools that can be leveraged to align KM initiatives with organizational goals. The chapter also examines the role of SKM in creating value-based healthcare and presents practical guidelines for developing a KM strategy that enhances organizational performance and competitive advantage.

The chapter is structured as follows. The first section examines generic strategic frameworks that can be applied to KM in healthcare. The second section discusses industry analysis techniques that help organizations understand the external knowledge environment. The third section explores internal analysis methods, focusing on Porter's Value Chain model and its application to KM. The fourth section introduces the Fishbone Analysis and McFarlan's Strategic Grid as tools for diagnosing KM challenges and opportunities. The fifth section provides guidance on designing a

KM strategy that aligns with organizational objectives. The sixth section examines the role of strategic KM in advancing value-based healthcare. The chapter concludes with a summary of key insights and implications for healthcare professionals.

GENERIC STRATEGIES

Porter's Generic Strategies and KM

Porter's (1985) generic strategies framework provides a useful starting point for examining strategic KM in healthcare (refer table 9.1). According to Porter, organizations can pursue three generic strategies to achieve competitive advantage namely cost leadership, differentiation, and focus. Each of these strategies has various implications for KM priorities and practices in healthcare organizations.

Cost Leadership Strategy

A cost leadership strategy in healthcare focuses on delivering services at a lower cost than competitors while maintaining acceptable quality standards. For organizations pursuing this strategy, KM initiatives should prioritize operational efficiency, standardization, and process optimization (Jalundhwala & Londhe, 2023).

KM can support cost leadership by ensuring that clinical and administrative processes are clearly documented and standardized, thereby minimizing inconsistencies and reducing operational waste. By identifying and disseminating best practices across the organization, KM enables healthcare providers to replicate cost-effective methods that enhance efficiency without compromising quality. Additionally, knowledge-sharing platforms create a structured environment for exchanging cost-saving ideas and innovations, allowing teams to learn from successful initiatives and avoid redundant efforts. The integration of performance analytics further strengthens cost leadership by leveraging data-driven insights to pinpoint inefficiencies, optimize workflows, and continuously refine operational strategies for sustainable cost reduction.

Mayo Clinic's implementation of a KM system exemplifies an effective approach to optimizing patient care. By standardizing clinical pathways and integrating knowledge into the clinical workflow, the institution has improved the accessibility and application of best practices. The AskMayoExpert system serves as a point-of-care resource, ensuring that clinicians have immediate access to vetted clinical knowledge, thereby enhancing decision-making and reducing variability in care delivery. As the program evolves, Mayo Clinic continues to explore models for improving knowledge representation and integration into electronic health records, aiming to support personalized and evidence-based patient care (Shellum et al., 2017).

Differentiation Strategy

A differentiation strategy involves offering unique features or services that create distinctive value for patients and other stakeholders. In healthcare, differentiation focusses on clinical excellence, patient experience, or innovative service models. KM initiatives aligned with a differentiation strategy should focus on innovation, expertise development, and personalized care delivery (Pflugfelder, 2021).

Innovation management in healthcare relies on capturing and utilizing knowledge to develop novel clinical approaches and service models that improve patient care and operational efficiency. By establishing expertise directories, organizations can systematically map and connect specialized knowledge across departments, ensuring that the right expertise is accessible when needed. Additionally, collecting and analyzing patient insights allows healthcare providers to understand patient preferences and tailor services, accordingly, fostering a more patient-centered approach. Collaborative research platforms further enhance innovation by facilitating knowledge exchange with academic partners, enabling healthcare institutions to stay at the forefront of clinical advancements and evidence-based practices.

Cleveland Clinic's Knowledge Program exemplifies this approach by integrating patient-reported outcomes with clinical data to personalize care delivery. Through systematic collection and analysis of patient feedback alongside clinical metrics, the program has generated valuable insights into patient needs and preferences. This comprehensive data-driven strategy has strengthened Cleveland Clinic's reputation for excellence in patient-centered care (Peeters et al., 2019).

Focus Strategy

A focus strategy involves targeting specific market segments or clinical functions, either with a cost leadership approach (focus on cost) or a differentiation approach (focus on differentiation). KM initiatives aligned with a focus strategy should emphasize specialized knowledge development and community engagement (Kothari et al., 2011).

Developing specialized knowledge repositories allows healthcare organizations to build comprehensive databases focused on specific medical conditions or patient populations, ensuring that relevant expertise and information are readily available. Supporting communities of practice fosters collaboration among specialists in targeted clinical areas, enabling the exchange of insights and best practices that drive continuous improvement. Additionally, patient education systems play a crucial role in delivering tailored educational resources that empower individuals to better manage their health and make informed decisions. Targeted research collaborations further strengthen knowledge development by fostering partnerships with specialized research institutions, advancing medical understanding and the application of innovative healthcare solutions.

Table 1. Alignment of Porter's generic strategies with KM priorities in healthcare

Generic Strategy	KM Focus	Key KM Initiatives
Cost Leadership	Efficiency and standardization	Process classification Best practice dissemination Knowledge sharing platforms Performance analytics
Differentiation	Innovation and expertise	Innovation management Expertise directories Patient insights collection Collaborative research platforms
Focus	Specialized knowledge and community engagement	Specialized knowledge repositories Community of practice support Patient education systems Targeted research collaborations

Resource-Based View (RBV) and KM

While Porter's framework focuses on positioning within the external environment, the RBV emphasizes the role of unique internal resources and capabilities in achieving competitive advantage (Gellweiler, 2018). In the context of healthcare, knowledge represents a critical resource that can be leveraged for strategic advantage (Evans et al., 2015). According to the RBV, resources must be valuable, rare, inimitable, and non-substitutable (VRIN) to provide sustainable competitive advantage as discussed below.

Clinical expertise and specialized knowledge are valuable assets that directly impact patient outcomes and overall organizational performance by enhancing decision-making and care quality. The rarity of such knowledge, particularly in specialized medical fields, creates a competitive advantage, as not all healthcare institutions have access to the same level of expertise or unique practice patterns. Additionally, the tacit knowledge gained through clinical experience is inimitable, making it difficult for competitors to replicate the depth of understanding that comes from years of hands-on practice. While technology can support and enhance medical decision-making, it remains non-substitutable, as it cannot fully replace the critical thinking, intuition, and personalized judgment that healthcare professionals bring to patient care.

KM strategies informed by the RBV focus on identifying, developing, and protecting knowledge assets that own VRIN characteristics (Corbo et al., 2023). Intermountain Healthcare's Advanced Training Program in quality improvement methodology exemplifies this approach. By systematically developing and disseminating proprietary knowledge in clinical process improvement, Intermountain has created a distinctive capability that contributes to its superior clinical and financial performance (Hulse et al., 2012).

Dynamic Capabilities and KM

The dynamic capabilities framework extends the RBV by emphasizing an organization's ability to adapt its resources in response to changing environments (Evans et al., 2015). This perspective is particularly relevant and important in healthcare, where rapid technological advancement and policy changes create a constantly evolving landscape (Bekbolatova et al., 2024).

From a KM perspective, dynamic capabilities play a crucial role in ensuring that healthcare organizations remain responsive to change. Knowledge sensing involves continuously identifying emerging trends, new medical research, and technological advancements that could impact healthcare practices. Once relevant knowledge is identified, knowledge seizing focuses on acquiring and integrating it into the organization's processes, ensuring that innovations and best practices are effectively adopted. Finally, knowledge reconfiguring enables organizations to adapt and restructure their existing knowledge assets, ensuring they remain relevant and applicable in addressing new challenges or opportunities in an evolving healthcare landscape.

Organizations with strong dynamic capabilities in KM can respond more effectively to disruptive changes in healthcare. Kaiser Permanente Northern California's Advance Alert Monitor (AAM) program exemplifies this capability by leveraging predictive analytics and virtual clinical monitoring to enhance early detection and intervention for at-risk patients. During the COVID-19 pandemic, Kaiser Permanente demonstrated rapid adaptation by integrating predictive modelling with telehealth strategies, ensuring continuity of care while reducing in-hospital deterioration. By systematically identifying high-risk patients and optimizing care pathways, the AAM program contributed to improved patient outcomes, reduced mortality rates, and enhanced hospital safety through a proactive and scalable knowledge-driven approach (Martinez et al., 2022).

INDUSTRY ANALYSIS

Understanding the external knowledge environment is vital for developing an effective KM strategy. Several analytical tools can help healthcare organizations assess the knowledge environment and identify strategic opportunities and threats.

Porter's Five Forces Analysis

Porter's Five Forces framework (Porter, 1980) provides a structured approach for analyzing competitive forces in an industry. This framework helps organizations understand the dynamics that influence knowledge acquisition, utilization, and protection in the healthcare knowledge environment (Refer Figure 1).

Figure 1. Porter's five forces applied to the healthcare knowledge environment

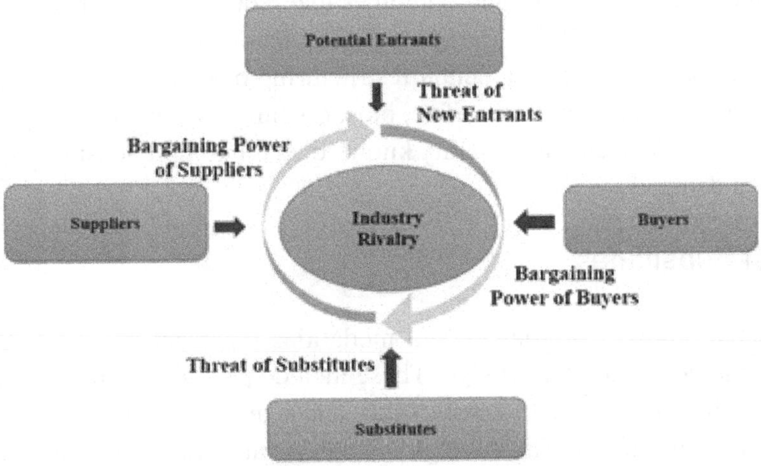

Threat of New Entrants

The healthcare knowledge has become increasingly accessible to new entrants, particularly technology companies with expertise in data analytics and artificial intelligence. Organizations like Google Health, Apple, and Amazon have entered the healthcare space, bringing advanced KM capabilities that competitive challenges and collaborative opportunities for traditional healthcare providers (O'Hara et al., 2017).

Bargaining Power of Suppliers

In the healthcare knowledge context, suppliers could be knowledge content providers (e.g., medical publishers, clinical decision support vendors), knowledge platform providers (e.g., electronic health record vendors), and knowledge service providers (e.g., consulting firms, education providers). The consolidation of these suppliers, particularly EHR vendors, has increased their bargaining power in recent years (O'Hara et al., 2017).

Bargaining Power of Buyers

Patients, insurers, and government payers could be identified as "buyers" who increasingly demand greater transparency, quality, and value in healthcare services. The rise of value-based payment models has expanded this bargaining power, with advances quality metrics and outcomes that require modern KM capabilities (Ndzabukelwako et al., 2024). The strategic implication extends beyond improved reporting mechanisms to fundamental restructuring of knowledge flows for value demonstration. Despite this imperative, most organizations maintain separate systems for clinical, financial, and quality knowledge, for comprehensive responses to value-based demands.

Threat of Substitutes

Alternative knowledge sources and modalities represent potential substitutes for traditional healthcare knowledge. These include patient-generated health data, artificial intelligence-based diagnostic tools, and health information services (Ndzabukelwako et al., 2024). The strategic challenge involves both competition and integration of these emerging knowledge sources. Organizations developing capabilities to evaluate and integrate alternative knowledge sources will likely establish significant competitive advantages.

Competitive Rivalry

Competition among healthcare organizations increasingly centers on knowledge-based capabilities, including clinical excellence, innovation, and patient engagement. Organizations that effectively leverage their knowledge assets often achieve competitive advantages in quality, efficiency, and market differentiation (Milstein et al., 2022). Healthcare executives frequently struggle with determining appropriate boundaries, improving community wide care while protecting proprietary knowledge assets that discuss competitive advantage. Few organizations have successfully resolved this fundamental tension.

PESTEL Analysis for Healthcare Knowledge Environment

The PESTEL framework examines the broader macro-environmental factors that impact an organization's strategic context. Applied to the healthcare knowledge environment, this analysis helps organizations anticipate external changes that may affect KM priorities and approaches. Refer Table 2.

Table 2. PESTEL analysis for healthcare KM

Factor	Key Trends and Developments	KM Implications
Political	Healthcare policy reforms Regulatory frameworks for data sharing Public health priorities	Alignment of knowledge strategy with policy directives Knowledge compliance systems for regulatory requirements Knowledge support for public health initiatives
Economic	Value-based payment models Healthcare cost containment pressures Investment in digital health	Knowledge systems for cost-effectiveness analysis ROI measurement for knowledge initiatives Strategic partnerships for knowledge technology investment
Social	Patient engagement and empowerment Health literacy challenges Aging population and chronic disease burden	Patient-centered knowledge design Accessible knowledge formats for diverse populations Knowledge support for chronic care management
Technological	Artificial intelligence and machine learning Interoperability standards Mobile health technologies	AI integration in knowledge systems Knowledge exchange architectures Mobile knowledge delivery platforms
Environmental	Sustainability in healthcare Climate-related health challenges Resource constraints	Knowledge sharing for sustainable practices Environmental health knowledge development Resource-efficient knowledge technologies
Legal	Data privacy regulations Intellectual property protection Professional liability concerns	Privacy-by-design in knowledge systems Knowledge asset protection strategies Evidence-based practice documentation

Knowledge SWOT Analysis

A Knowledge SWOT (Strengths, Weaknesses, Opportunities, Threats) analysis provides a focused assessment of an organization's knowledge position relative to its strategic context. This analysis helps healthcare organizations identify knowledge-related strengths to leverage, weaknesses to address, opportunities to pursue, and threats to mitigate (Benzaghta et al., 2021). A knowledge SWOT analysis should address the respective questions as illustrated in Figure 2.

Figure 2. Knowledge SWOT analysis for a hypothetical healthcare organization

Strengths	**Weaknesses**
What unique knowledge assets does the organization possess?	What knowledge gaps exist within the organization?
In which clinical domains does the organization have superior knowledge?	Where do knowledge silos or barriers to knowledge sharing occur?
What KM processes are particularly effective?	What KM capabilities are underdeveloped?
Opportunities	**Threats**
What emerging knowledge domains could the organization develop?	What competitive knowledge advantages might erode?
How could external partnerships enhance knowledge capabilities?	How might technological changes disrupt current knowledge models?
What patient needs could be addressed through improved knowledge utilization?	What compliance risks exist in KM practices?

The Knowledge SWOT analysis serves as a bridge between external industry analysis and internal organizational assessment, providing a comprehensive view of the strategic knowledge landscape.

INTERNAL ANALYSIS PORTER'S VALUE CHAIN MODEL

Porter's Value Chain model (Porter, 1985) provides a systematic approach for analyzing internal organizational activities and their contribution to value creation. Applied to KM in healthcare, this model helps organizations identify where and how knowledge adds value across the care delivery process.

The Knowledge Value Chain in Healthcare

The traditional value chain model consists of two types of activities namely primary activities (directly related to production and delivery) and support activities (enabling and enhancing primary activities). In healthcare, the primary activities include patient acquisition, assessment and diagnosis, treatment planning, care delivery, and post-discharge management. The support activities include infrastructure management, human resource management, technology development, and procurement (Ali & Kannan, 2022). Knowledge plays a critical role across all these activities, creating what can be named a "knowledge value chain" (Faroukhi et al., 2020). Figure 9.3 illustrates the knowledge value chain in healthcare and identifies key knowledge processes at each stage.

Figure 3. Knowledge value chain in healthcare

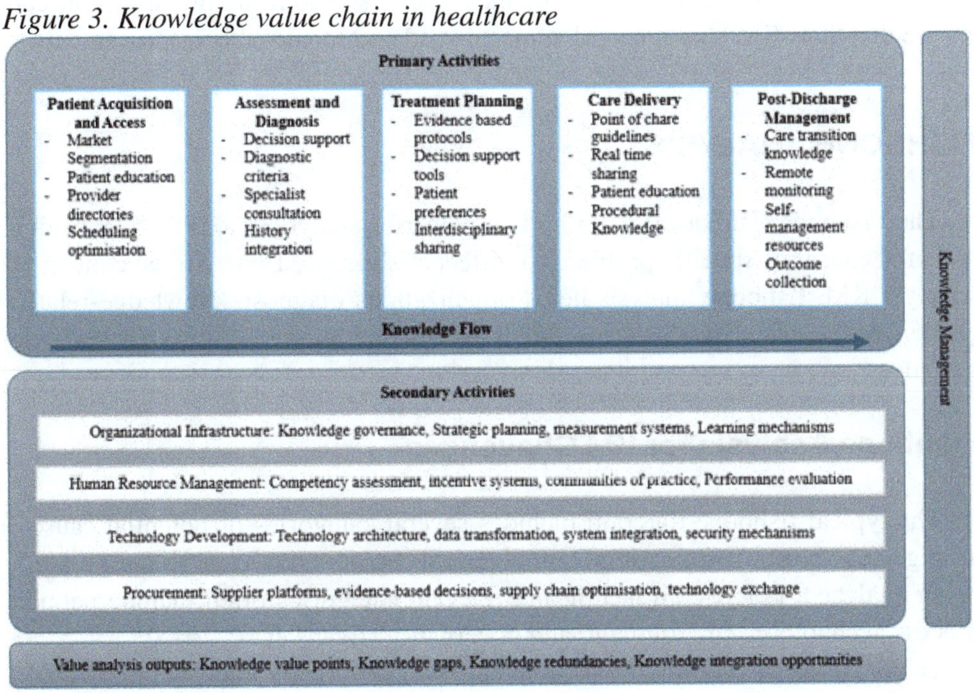

Knowledge Margin Analysis

Porter's Value Chain model includes the concept of "margin" the difference between the total value created and the cost of performing value activities. Knowledge margin represents the additional value generated through effective KM relative to the investment in knowledge processes (Cassia et al., 2020).

Knowledge margin can be evaluated using several key metrics that assess the effectiveness and value of KM initiatives in healthcare. Knowledge Return on Investment (ROI) measures the overall value generated by knowledge initiatives compared to the costs of implementing and maintaining KM systems, ensuring that investments lead to tangible improvements in efficiency and patient care. The Knowledge Efficiency Ratio assesses how effectively knowledge is utilized to drive operational improvements, comparing enhancements in workflow, decision-making, and resource optimization against the expenses associated with knowledge processes. Additionally, Knowledge Quality Impact evaluates the influence of knowledge enhancements on clinical outcomes, analyzing how improved access to and application of knowledge contribute to better patient care, reduced errors, and overall healthcare performance. By leveraging these metrics, healthcare organizations can ensure that their KM strategies drive meaningful and measurable improvements.

FISHBONE ANALYSIS

The Fishbone (Ishikawa) diagram is a powerful analytical tool for identifying potential causes of specific problems or effects (Ishikawa, 1990). In the context of strategic KM, fishbone analysis helps organizations diagnose knowledge-related challenges and identify root causes that need to be addressed through strategic initiatives.

Fishbone Analysis for KM Challenges

The typical fishbone diagram includes several categories of potential causes, arranged as "bones" originating from a central "spine" that leads to the problem being analyzed. For KM in healthcare, relevant categories often include people, process, technology, environment, and governance (Evans et al., 2015).

Table 3 presents common KM challenges in healthcare and potential causes identified through fishbone analysis.

Table 3. Fishbone analysis for common KM challenges in healthcare

Knowledge Challenge	People Factors	Process Factors	Technology Factors	Environmental Factors	Governance Factors
Inadequate knowledge sharing across departments	- Siloed professional cultures - Lack of collaboration incentives - Territorial behaviors	- Undefined sharing protocols - Inefficient communication channels - Inadequate transfer mechanisms	- Incompatible systems - Poor interface design - Limited accessibility	- Physical separation - Time constraints - Competitive culture	- Unclear knowledge ownership - Absence of sharing policies - Lack of leadership emphasis
Limited capture of tacit knowledge	- Expert time constraints - Resistance to codification - Limited documentation skills	- Inadequate capture mechanisms - Informal knowledge processes - Insufficient reflection time	- Cumbersome documentation tools - Limited multimedia capabilities - Poor integration with workflow	- Productivity pressures - Interruption-rich setting - Crisis-oriented culture	- No tacit knowledge valuation - Missing capture incentives - Lack of knowledge standards
Poor knowledge retrieval at point of care	- Limited search skills - Time pressure - Preference for social knowledge sources	- Workflow interruptions - Complex access procedures - Insufficient content curation	- Poor search functionality - Slow system response - Inadequate mobile access	- Noisy environment - Multiple competing priorities - Patient interruptions	- No retrieval standards - Limited training requirements - Unclear knowledge authority

Strategic Applications of Fishbone Analysis

Healthcare organizations can leverage fishbone analysis to enhance strategic KM by systematically identifying and addressing key challenges. Organizations can focus on root causes that have the most significant impact on KM inefficiencies, ensuring that strategic initiatives are targeted and effective. The structured nature of fishbone analysis supports the development of comprehensive solutions, as it considers multiple contributing factors such as people, processes, technology, and governance allowing for a holistic approach to overcoming knowledge-related barriers. Additionally, the visual representation of knowledge challenges facilitates stakeholder alignment, fostering a shared understanding among healthcare professionals, administrators, and policymakers, thereby ensuring commitment to collective problem-solving. Finally, fishbone analysis serves as a valuable tool for measuring progress, enabling organizations to establish baseline assessments and track improvements over time, ensuring that interventions lead to meaningful and sustainable advancements in KM effectiveness.

Johns Hopkins Medicine employed a gap analysis tool to identify barriers to evidence-based practice adoption across its healthcare system. The analysis revealed that while significant investments had been made in technology, the primary obstacles were related to people and processes particularly clinical time constraints and challenges with workflow integration. These findings led to a strategic shift toward embedding knowledge directly into clinical workflows rather than merely improving access to information. As a result, Johns Hopkins developed integrated clinical pathways that seamlessly incorporate evidence-based practices into routine care (Golden et al., 2017).

MCFARLAN'S STRATEGIC GRID

McFarlan's Strategic Grid (McFarlan, 1984) provides a framework for assessing the strategic importance of different information systems and technologies to an organization. This model can be adapted for KM, helping healthcare organizations categorize and prioritize knowledge initiatives based on their strategic significance.

McFarlan's Grid for KM

The adapted McFarlan's Grid for KM classifies knowledge systems along two dimensions:

1. Current Strategic Impact: The extent to which existing knowledge systems contribute to organizational strategic objectives.
2. Future Strategic Impact: The potential of emerging knowledge initiatives to transform organizational capabilities and competitive position.

These dimensions create four quadrants, each with distinct strategic implications (refer Figure 4)

Figure 4. McFarlan's strategic grid for KM

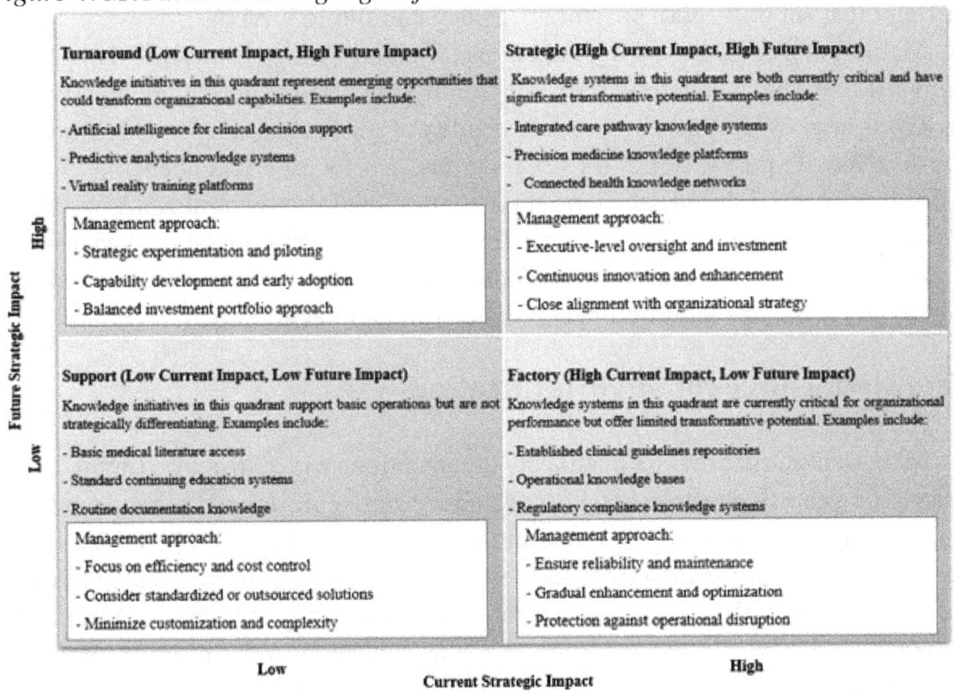

Strategic Knowledge Portfolio Management

McFarlan's Grid helps healthcare organizations create a well-balanced portfolio of knowledge initiatives by addressing both immediate operational needs and long-term strategic goals. Core Knowledge Systems, positioned in the Factory Quadrant, ensure that essential knowledge resources are consistently available for day-to-day operations, supporting efficiency and reliability in healthcare delivery. Transformative Knowledge Initiatives, found in the Turnaround Quadrant, focus on developing new capabilities that can provide a future competitive advantage, helping organizations adapt to evolving industry trends and technological advancements. In the Strategic Quadrant, Strategic Knowledge Platforms encompass knowledge systems that deliver sustained value by integrating both current and future knowledge requirements, allowing for seamless innovation and scalability. Lastly, the Support Quadrant includes foundational KM functions that, while not directly strategic, are necessary to maintain an effective knowledge infrastructure. By balancing investments across these quadrants, healthcare organizations can ensure a robust, forward-thinking KM strategy that supports both present efficiency and future growth.

Maintaining basic knowledge functions efficiently. Mayo Clinic's KM approach demonstrates a balanced strategy by integrating structured clinical guidelines within its KM system. While ensuring the accessibility of vetted clinical knowledge through the AskMayoExpert system, Mayo Clinic has also focused on enhancing clinical workflow integration and decision support. Additionally, its centralized Knowledge Content Management System streamlines knowledge storage, updating, and dissemination, ensuring that foundational medical knowledge remains standardized and readily available to support clinical practice (Shellum et al., 2017).

Knowledge Investment Prioritization

McFarlan's Grid provides healthcare organizations with a structured approach to prioritizing knowledge investments based on their strategic importance and potential impact. Strategic investments, found in the Strategic Quadrant, require significant funding and executive attention, as they contribute both immediate benefits and long-term competitive advantage. Selective investments, placed in the Turnaround Quadrant, demand a more measured approach, balancing the promise of future innovation with existing resource limitations, ensuring that investments are phased appropriately. In the Factory Quadrant, efficiency investments focus on maintaining reliable knowledge systems, optimizing processes, and making incremental improvements rather than undertaking large-scale transformations. Finally, standardization investments, categorized under the Support Quadrant, prioritize cost-efficiency, system consolidation, and ensuring a consistent knowledge infrastructure across the organization. By carefully distributing resources across these investment types, healthcare organizations can develop a sustainable and forward-looking KM strategy.

DESIGNING A KM STRATEGY

Developing an effective KM (KM) strategy requires a systematic approach that aligns knowledge initiatives with organizational objectives, addresses key knowledge challenges, and leverages strategic opportunities. This section outlines a structured process for designing a KM strategy in healthcare organizations.

Strategic Knowledge Planning Process

The strategic knowledge planning process involves several interconnected stages that guide organizations from initial assessment to implementation and evaluation (see Figure 9.6).

Figure 5. Strategic knowledge planning process

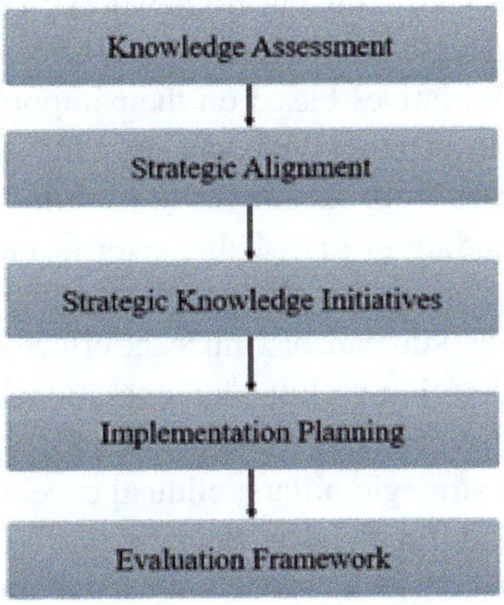

Knowledge Assessment

The knowledge assessment phase is a crucial step in understanding an organization's current knowledge landscape, serving as the foundation for an effective KM strategy. It begins with a knowledge asset inventory, where both explicit and tacit knowledge resources are catalogued based on their location, format, quality, and accessibility, ensuring a clear picture of available intellectual assets. Knowledge flow analysis then examines how knowledge moves within the organization whether smoothly or with barriers between individuals, departments, and external stakeholders. Identifying knowledge gaps is another critical component, as it helps pinpoint areas where essential knowledge is lacking and needs to be developed to support strategic goals. Finally, a knowledge SWOT analysis provides a structured evaluation of the organization's knowledge strengths, weaknesses, opportunities, and threats, offering valuable insights for aligning knowledge initiatives with broader organizational objectives. Through this structured assessment, healthcare organizations can make informed decisions about how to enhance their KM practices.

Strategic Alignment

The strategic alignment phase is essential for ensuring that KM (KM) initiatives directly contribute to an organization's broader strategic goals. This process begins with strategic knowledge mapping, where organizations identify the specific knowledge needs required to support key objectives and initiatives. Once these requirements are clear, knowledge value analysis helps determine where knowledge can provide the most significant impact, often using frameworks like value chain analysis to pinpoint areas of high strategic value. Knowledge prioritization follows, ranking knowledge domains and initiatives based on their importance, with tools such as McFarlan's Grid guiding investment decisions. Finally, stakeholder alignment ensures that all relevant parties from executives to frontline healthcare professionals share a common understanding of knowledge priorities and remain committed to their successful implementation. By aligning KM initiatives with organizational strategy, healthcare organizations can maximize the effectiveness of their knowledge resources, driving better decision-making and operational excellence.

Intermountain Healthcare's strategic alignment process mapped knowledge requirements across its four strategic pillars: clinical excellence, patient experience, physician engagement, and financial sustainability. This revealed that clinical knowledge integration served as a keystone knowledge domain supporting all four pillars (Nuti & Vainieri, 2014).

Strategic Knowledge Initiatives

Building on the knowledge assessment and strategic alignment phases, organizations can implement targeted strategic knowledge initiatives to enhance their KM capabilities (refer table 9.4). Knowledge generation focuses on creating or acquiring new insights through research collaborations and innovation programs, ensuring that organizations stay at the forefront of clinical advancements. Once knowledge is developed, knowledge codification plays a crucial role in capturing and structuring it into usable formats, such as clinical guidelines or expertise mapping, making it accessible and standardized across the organization. To ensure that knowledge reaches the right people, knowledge transfer facilitates sharing through mechanisms like communities of practice and learning networks, promoting collective learning and collaboration. The next step, knowledge application, ensures that knowledge is effectively integrated into daily practice through clinical decision support tools and just-in-time learning strategies, improving patient care and operational efficiency. Finally, knowledge infrastructure focuses on strengthening foundational systems, such as governance frameworks and technology platforms, to support sustainable KM efforts. By implementing these strategic initiatives, healthcare organizations can optimize knowledge utilization and drive continuous improvement in healthcare delivery.

Table 4. Strategic knowledge initiatives by process and objective

Knowledge Process	Clinical Excellence	Patient Experience	Operational Efficiency	Innovation
Generation	Clinical research collaborations	Patient needs assessment	Process improvement research	Innovation labs
	Quality improvement studies	Experience mapping research	Workflow optimization studies	Research partnerships
	Clinical registry development	Patient journey analysis	Benchmarking analyses	Open innovation platforms
Codification	Clinical guideline development	Patient education content development	Standard operating procedure development	Innovation methodology documentation
	Best practice documentation	Experience standard documentation	Process documentation	Design thinking playbooks
	Rare case knowledge bases	Service script development	Role knowledge mapping	Lessons learned repositories
Transfer	Clinical communities of practice	Patient story sharing platforms	Process knowledge sharing forums	Innovation showcase events
	Case-based learning forums	Experience insight dissemination	Best practice diffusion networks	Cross-functional innovation teams
	Expert consultation networks	Patient education delivery systems	Performance transparency systems	Idea marketplaces
Application	Clinical decision support	Personalized knowledge delivery	Performance dashboard integration	Rapid prototyping systems
	Point-of-care knowledge tools	Just-in-time patient education	Process guidance systems	Innovation testing platforms
	Embedded evidence alerts	Preference-informed care tools	Operational decision support	Simulation environments
Infrastructure	Clinical knowledge governance	Patient feedback systems	Performance knowledge analytics	Innovation management systems
	Evidence evaluation systems	Experience measurement platforms	Process mining capabilities	Collaboration platforms
	Quality measurement frameworks	Voice-of-patient integration	Operational knowledge repositories	Idea evaluation frameworks

Implementation Planning

This phase focuses on translating SKM initiatives into concrete, actionable plans that can be effectively implemented within an organization. It begins with developing an initiative roadmap, which lays out a clear sequence of initiatives over time, complete with defined milestones to track progress. Once the roadmap is established, resource allocation ensures that the necessary human, financial, and technological resources are identified and distributed appropriately to support the initiatives. A strong governance structure is then put in place to define oversight mechanisms, establish decision-making authority, and ensure accountability across teams. Finally, a change management plan is essential for addressing cultural, behavioral, and process shifts that may arise during implementation. By proactively managing these changes, organizations can foster a smooth transition, ensuring that KM initiatives are successfully adopted and contribute to long-term strategic success.

Mayo Clinic's three-year KM roadmap began with foundational infrastructure initiatives, followed by clinical knowledge integration projects, and concluded with advanced analytical capabilities. This phased approach ensured each stage built upon the last (Pike & Mongan, 2014).

Evaluation Framework

The evaluation framework provides a structured approach for assessing the effectiveness of KM initiatives and their alignment with strategic objectives. Knowledge metrics serve as key indicators to measure knowledge-related processes, such as the rate of knowledge creation and the frequency of utilization, ensuring that knowledge is actively generated and applied. To gauge the real impact of KM initiatives, outcome metrics assess their contribution to broader organizational goals, including improvements in clinical quality, patient experience, and operational efficiency. The effectiveness of these metrics is supported by evaluation methods, which employ both quantitative and qualitative approaches to analyze the data, offering a comprehensive understanding of KM performance. Finally, feedback mechanisms ensure that evaluation insights are not only recorded but actively used to refine and adapt KM initiatives, creating a continuous cycle of learning and improvement. Through this structured evaluation process, healthcare organizations can optimize their knowledge strategies and maximize their impact on patient care and organizational success.

Mayo Clinic's KM roadmap followed a structured, phased approach, beginning with the establishment of a foundational knowledge infrastructure. This was followed by initiatives to integrate clinical knowledge into workflows through systems like AskMayoExpert and the Knowledge Content Management System. The final phase focused on enhancing decision support capabilities and exploring advanced knowledge modelling techniques to improve clinical practice. This approach ensured each stage built upon the previous one, fostering a cohesive and scalable KM strategy (Shellum et al., 2017).

Knowledge Strategy Development Approaches

Organizations can adopt different approaches to developing a KM strategy, depending on their context, culture, and capabilities.

Top-Down Approach

A top-down approach to KM emphasizes a centralized strategy where leadership drives decision-making and implementation across the organization. In this model, executive leadership plays a key role in defining the KM vision, setting priorities, and ensuring alignment with organizational goals. The approach begins with an organization-wide knowledge assessment and planning process, ensuring that KM initiatives are structured, data-driven, and strategically focused. To maintain consistency and efficiency, standardized KM processes and systems are implemented across departments, creating a unified framework for knowledge sharing, storage, and utilization. Additionally, a centralized governance structure oversees KM initiatives, allocating resources effectively and ensuring accountability at all levels. While this approach ensures strong alignment and consistency, it requires careful management to encourage engagement from various teams and prevent rigidity in adapting to local knowledge needs.

While alignment and standardization are strengths, challenges may arise around engagement and adaptability to local needs.

Bottom-Up Approach

A bottom-up approach to KM (KM) focuses on empowering front-line teams to drive knowledge initiatives based on their specific needs and experiences. This model begins with a local knowledge needs assessment and planning process, where departments and teams identify gaps, challenges, and opportunities within their own operational context. Instead of relying on a centralized directive, department-level KM initiatives emerge organically, allowing teams to develop and implement solutions that are most relevant to their work. Governance in this approach is decentralized, giving departments more control over their knowledge processes while still aligning with broader organizational objectives. Additionally, network-based sharing and coordination play a crucial role, fostering collaboration between teams and enabling best practices to spread across the organization. While this approach encourages innovation, engagement, and responsiveness to on-the-ground challenges, it requires mechanisms to ensure alignment with overall organizational strategy and to prevent knowledge silos.

Strengths include engagement, relevance, and contextual adaptation, though strategic alignment and efficiency can be challenging.

Hybrid Approach

Most healthcare organizations adopt a hybrid approach to KM, integrating both top-down and bottom-up elements to balance strategic alignment with local adaptability. This approach begins with an organization-wide KM framework and priorities, ensuring that overarching goals, standards, and best practices guide knowledge initiatives across the institution. At the same time, local adaptation of KM initiatives allows individual departments and teams to tailor KM strategies to their specific operational needs, fostering flexibility and responsiveness. Governance follows a federated model, where centralized oversight provides strategic direction while local units retain autonomy to manage knowledge resources effectively. Additionally, shared infrastructure with local customization ensures that all teams have access to common KM tools and systems, while still allowing for adjustments that meet their unique requirements. By combining the strengths of both centralized coordination and decentralized innovation, the hybrid approach promotes efficiency, engagement, and continuous learning across the organization.

Partners HealthCare's hybrid approach established a central KM Office for strategic direction and infrastructure while empowering clinical departments to develop specialized initiatives aligned with the overall framework (Davenport & Glaser, 2018).

Knowledge Strategy Archetypes

Healthcare organizations often adopt one of several knowledge strategy archetypes, based on their strategic position, organizational characteristics, and knowledge priorities.

Codification Strategy

A codification strategy in KM emphasizes the systematic documentation, organization, and dissemination of explicit knowledge, making it easily accessible across the organization. This approach relies on comprehensive knowledge repositories, ensuring that critical information, such as clinical guidelines, best practices, and research findings, is stored in a structured and retrievable format. To maintain consistency and reliability, standardized documentation processes are implemented, ensuring that knowledge is captured in a uniform manner across different teams and departments. Additionally, advanced search and retrieval capabilities enable healthcare professionals to quickly locate relevant information, reducing inefficiencies and enhancing decision-making. Finally, knowledge quality assurance mechanisms ensure that the information remains accurate, up-to-date, and reliable, reinforcing trust in the KM system. This structured approach is particularly beneficial in healthcare settings where precise, evidence-based knowledge is essential for patient care and operational excellence.

Ideal for organizations focusing on standardization, reliability, and scalability.

Personalization Strategy

A personalization strategy in KM focuses on person-to-person knowledge sharing, ensuring that tacit knowledge gained through experience and expertise is effectively transferred across the organization. This approach invests in expert directories and connection platforms, enabling employees to identify and connect with subject matter experts for guidance and collaboration. Communities of practice and professional networks further enhance knowledge exchange by fostering discussions, shared problem-solving, and innovation within specialized areas. Mentoring and shadowing programs play a crucial role in transferring deep, experience-based knowledge from senior professionals to newer team members, ensuring continuity and skill development. Additionally, collaborative workspaces create environments where teams can engage in real-time discussions, brainstorming, and knowledge co-creation. By emphasizing direct human interaction and experiential learning, a personalization strategy is particularly effective in complex healthcare environments where contextual decision-making, expert judgment, and adaptability are critical.

Ideal for organizations emphasizing innovation, complex problem-solving, and personalized service.

Balanced Strategy

A balanced KM strategy combines elements of both codification and personalization, ensuring that knowledge is efficiently documented while also fostering direct human interaction for complex decision-making. Codification is applied to established clinical knowledge and standardized processes, where structured documentation, repositories, and guidelines help ensure consistency and accessibility across the organization. At the same time, personalization is emphasized for complex cases, emerging knowledge domains, and innovation, where direct collaboration, mentoring, and expertise-sharing are essential for nuanced problem-solving. To support this dual approach, integrated and flexible KM platforms provide a seamless environment where structured knowledge storage coexists with dynamic knowledge-sharing tools, ensuring that both explicit and tacit knowledge are leveraged effectively. This strategy enables healthcare organizations to maintain operational efficiency while fostering innovation and adaptability in an evolving medical landscape.

Mayo Clinic's balanced approach standardizes established clinical knowledge while enabling personalized application in complex and innovative domains (Shellum et al., 2017) (refer Table 5).

Table 5. Comparison of knowledge strategy archetypes

Dimension	Codification Strategy	Personalization Strategy	Balanced Strategy
Primary Knowledge Focus	Explicit knowledge	Tacit knowledge	Both explicit and tacit knowledge
Knowledge Assets	Documents, databases, protocols	Expertise, relationships, experience	Complementary explicit and tacit assets
Technology Emphasis	Content management, search, retrieval	Collaboration, connection, communication	Integrated knowledge platforms
Organization Structure	Centralized knowledge functions	Distributed knowledge networks	Federated knowledge model
Culture Emphasis	Documentation, standardization, reuse	Sharing, collaboration, mentoring	Contextual knowledge approach
Measurement Focus	Content quantity, utilization, quality	Network connectivity, expertise location	Multi-dimensional knowledge metrics
Strategic Advantage	Efficiency, reliability, scalability	Innovation, adaptability, personalization	Comprehensive knowledge capabilities
Ideal Context	Stable, procedural knowledge domains	Dynamic, complex knowledge domains	Diverse knowledge ecosystem
Healthcare Example	Kaiser Permanente's clinical guideline system	Mayo Clinic's specialist consultation network	Cleveland Clinic's knowledge integration platform

VALUE BASED HEALTHCARE

Value-based healthcare represents a paradigm shift from volume-based to value-based incentives, with value defined as health outcomes achieved per dollar spent (Porter & Teisberg, 2006). Strategic KM plays a critical role in enabling this transformation, as value-based care requires strong approaches to capturing, analyzing, and utilizing knowledge about outcomes, processes, and costs.

The Value Equation in Healthcare

Porter's value equation defines value as health outcomes that matter to patients divided by the cost of delivering those outcomes (Porter, 2010). This seemingly simple equation has profound implications for KM, as both the numerator (outcomes) and denominator (costs) require high-level knowledge capabilities.

Outcome KM

Delivering value-based care relies on a well-structured KM system that provides comprehensive insights into various outcome dimensions. Clinical outcomes serve as the foundation, encompassing key indicators such as mortality rates, morbidity, complication rates, functional status, and disease control, all of which reflect the effectiveness of medical interventions. Equally important are patient-reported outcomes, which capture the patient's perspective on their symptoms, quality of life, satisfaction with care, and overall goal achievement, offering a more holistic view of healthcare success. Finally, process outcomes focus on the efficiency and quality of care delivery, including timeliness, coordination among providers, patient safety, and adherence to evidence-based medical practices. By systematically capturing and analyzing knowledge related to these outcomes, healthcare organizations can drive continuous improvement, enhance patient-centered care, and ensure that value-based care models achieve their intended impact.

Effective KM systems are essential for supporting value-based care by ensuring the systematic collection, integration, analysis, and utilization of outcome measures. To achieve this, organizations must first develop outcome measurement knowledge, which involves defining, collecting, and validating key healthcare outcomes to ensure accuracy and consistency in performance tracking. Once data is collected, outcome analysis knowledge plays a critical role in conducting risk-adjustment, benchmarking against industry standards, and identifying trends that offer insights into care effectiveness. Finally, outcome improvement knowledge focuses on translating these insights into actionable strategies for enhancing patient care, optimizing clinical workflows, and implementing continuous quality improvement initiatives. By leveraging these knowledge-driven capabilities, healthcare organizations can align their practices with value-based care principles, leading to better patient outcomes, increased efficiency, and sustainable healthcare improvements.

Cleveland Clinic's Knowledge Program exemplifies this approach, integrating clinical and patient-reported outcomes into a unified knowledge platform that supports value-based care initiatives (Peeters et al., 2019).

Cost KM

Managing the cost component of the value equation in healthcare requires sophisticated KM capabilities to ensure that financial resources are used efficiently without compromising quality. Cost measurement knowledge is essential for accurately tracking the full cycle of costs across an entire care episode, from initial consultation to post-treatment follow-up, allowing for a complete understanding of healthcare expenditures. Cost attribution knowledge enhances financial transparency by linking specific costs to conditions, medical processes, and patient outcomes, ensuring that organizations can pinpoint areas where expenses are highest and assess whether they align with value-based care goals. Lastly, cost optimization knowledge focuses on identifying opportunities to reduce costs while maintaining or even improving patient outcomes. This involves analyzing inefficiencies, reallocating resources effectively, and leveraging data-driven strategies to enhance financial sustainability. By integrating these knowledge-driven approaches, healthcare organizations can better manage costs, improve resource utilization, and enhance overall value in patient care.

Effective KM systems play a critical role in helping healthcare organizations understand their true costs and uncover opportunities for cost optimization while maintaining high-quality care. This begins with cost accounting knowledge, which ensures that costs are accurately captured and allocated across various departments, procedures, and patient care processes, providing a clear financial picture. Next, cost variation knowledge enables organizations to analyze cost drivers and identify inconsistencies or inefficiencies in spending, allowing for targeted interventions to improve financial efficiency. Lastly, cost-effectiveness knowledge focuses on evaluating the relationship between costs and patient outcomes, ensuring that financial decisions are aligned with value-based care principles. By integrating these capabilities, healthcare organizations can make data-driven financial decisions, optimize resource allocation, and achieve a sustainable balance between cost control and quality improvement.

Intermountain Healthcare's Activity-Based Costing system exemplifies an approach that aligns with strategies for managing medical practice variation by enhancing efficiency while maintaining or improving care quality. Similar to performance evaluation frameworks such as those implemented in the Tuscany Region, Intermountain's system provides detailed insights into the costs associated with specific clinical processes and their relationship to patient outcomes. By leveraging this knowledge, Intermountain has optimized resource allocation and improved performance across multiple clinical domains. This aligns with broader efforts in healthcare systems to balance cost containment with quality improvement through data-driven decision-making and standardized best practices (Nuti & Vainieri, 2014).

KM for Value-Based Care Models

Value-based care encompasses several specific delivery and payment models, each with distinct KM requirements.

Accountable Care Organizations (ACOs)

Accountable Care Organizations (ACOs) are networks of healthcare providers that share responsibility for delivering high-quality, cost-effective care to a defined population. To operate efficiently and meet value-based care objectives, ACOs require robust KM systems that support informed decision-making and coordination. Population KM enables organizations to systematically assess the health status, risks, and specific needs of their attributed patient population, ensuring that care strategies are data-driven and patient-centered. Network KM focuses on facilitating seamless information exchange across providers within the ACO, promoting collaboration, reducing redundancies, and improving care coordination. Additionally, performance KM is essential for continuously tracking, analyzing, and optimizing both quality and cost outcomes, allowing ACOs to refine their strategies, meet performance benchmarks, and drive sustainable improvements in healthcare delivery. By leveraging these KM capabilities, ACOs can enhance efficiency, improve patient outcomes, and achieve long-term success in value-based care models.

Bundled Payment Models

Bundled payment models streamline healthcare financing by providing a single payment for all services related to a specific episode of care, incentivizing providers to enhance efficiency while maintaining quality. To support this approach, KM systems must ensure effective tracking and utilization of healthcare data. Episode KM is essential for defining, monitoring, and analyzing care episodes, ensuring that providers have a clear understanding of the services included and the expected outcomes. Coordination KM facilitates seamless knowledge sharing across all providers involved in a patient's care journey, improving communication, reducing duplication of efforts, and ensuring continuity of care. Lastly, variation KM helps identify differences in treatment processes, resource utilization, and costs across providers, enabling organizations to pinpoint inefficiencies and standardize best practices. By integrating these KM capabilities, bundled payment models can drive cost-effectiveness while ensuring high-quality, patient-centered care.

The Baptist Health System's bundled payment knowledge platform exemplifies this approach, providing comprehensive knowledge about episode definitions, coordination requirements, and variation patterns. This platform has enabled Baptist to reduce costs by 20% while improving quality metrics across multiple bundled payment programs (Navathe et al., 2017).

Patient-Centered Medical Homes (PCMHs)

Patient-Centered Medical Homes (PCMHs) emphasize comprehensive, coordinated primary care with a whole-person approach, integrating physical, behavioral, and social health needs. To effectively support this model, KM systems must facilitate the seamless flow of information across various aspects of care. Comprehensive care knowledge ensures that data related to medical, mental health, and social determinants of health are integrated into a unified system, allowing providers to make well-informed, holistic care decisions. Coordination knowledge plays a crucial role in aligning care across the healthcare ecosystem, ensuring smooth collaboration between primary care providers, specialists, social services, and community resources. Additionally, engagement knowledge focuses on capturing and utilizing patient preferences, personal health goals, and self-management capabilities, empowering individuals to take an active role in their care. By leveraging these KM capabilities, PCMHs can enhance care continuity, patient satisfaction, and health outcomes, ultimately driving more effective and patient-centered primary care delivery.

Geisinger Health System's ProvenHealth Navigator knowledge platform exemplifies this approach, integrating comprehensive care knowledge, coordination tools, and patient engagement capabilities. This platform has enabled Geisinger to reduce hospital admissions and improve diabetes control rates (Pike & Mongan, 2014).

Knowledge-Enabled Value Transformation

Strategic KM can enable healthcare organizations to drive value-based transformation through several mechanisms.

Knowledge-Enabled Care Redesign

Value-based care necessitates a fundamental redesign of care delivery models, with KM serving as a key enabler in optimizing patient outcomes and healthcare efficiency. One critical approach is condition-based organization, which restructures care delivery around specific medical conditions or patient segments, ensuring that providers have access to integrated knowledge on best practices, expected outcomes, and cost-effectiveness for targeted treatment strategies. Care pathway integration further enhances this model by developing structured care pathways that guide patients through the entire care cycle, incorporating evidence-based knowledge on optimal treatment processes, critical decision points, and expected health outcomes.

Additionally, team-based knowledge models foster a culture of collaboration by enabling seamless knowledge sharing within multidisciplinary care teams, ensuring that insights from different specialties contribute to holistic and coordinated patient care. By embedding these KM strategies, healthcare organizations can enhance care quality, streamline operations, and drive sustainable improvements in value-based care delivery.

Knowledge-Enabled Performance Improvement

Value-based care relies on continuous improvement in both patient outcomes and cost efficiency, with KM playing a central role in driving systematic advancements. Outcome variation analysis enables organizations to identify discrepancies in patient outcomes across different providers, locations, or population groups, using data-driven insights to implement targeted improvements and reduce inconsistencies in care quality. Best practice transfer ensures that successful clinical methods and operational efficiencies are systematically documented and shared across the organization, allowing teams to adopt proven strategies and enhance overall performance. Additionally, the learning health system approach fosters a culture of continuous improvement by integrating knowledge from clinical care, research findings, and quality improvement initiatives, creating a dynamic feedback loop that refines care delivery over time. By embedding these knowledge-driven methodologies, healthcare organizations can optimize value-based care, ensuring better patient experiences, improved health outcomes, and sustainable cost management.

Intermountain Healthcare's Advanced Training Program in Healthcare Delivery Improvement exemplifies this approach, developing sophisticated knowledge capabilities for performance improvement across the organization. This program has enabled Intermountain to achieve top-decile performance on quality metrics while maintaining costs at 30% below peer institutions (Olsen et al., 2021).

Knowledge-Enabled Patient Engagement

Value-based care places a strong emphasis on active patient engagement, ensuring that individuals play a central role in decision-making, self-management, and outcome reporting. KM facilitates this engagement by providing shared decision-making knowledge, which equips both patients and clinicians with evidence-based information to support preference-sensitive healthcare choices, fostering informed discussions and personalized treatment plans. Additionally, self-management knowledge empowers patients with tailored educational resources, tools, and guidance that enable them to effectively manage their own health conditions, improving adherence to treatment and long-term well-being. Lastly, patient-reported outcome knowledge ensures that healthcare organizations systematically capture and utilize patient feedback on symptoms, quality of life, and treatment effectiveness, allowing providers to refine care strategies based on real-world patient experiences. By integrating these knowledge-driven capabilities, healthcare systems can enhance patient involvement, improve health outcomes, and drive the success of value-based care initiatives.

Strategic KM Maturity for Value-Based Healthcare

Organizations can assess their strategic KM maturity for value-based healthcare using a capability maturity model. Table 6 presents a maturity model for KM capabilities in value-based care.

Table 6. KM maturity model for value-based healthcare

Domain	Level 1: Initial	Level 2: Developing	Level 3: Defined	Level 4: Managed	Level 5: Optimizing
Outcome Knowledge	Limited outcome measures	Multiple outcome measures	Comprehensive measures	Patient-centered measures	Personalized measures
	Fragmented collection	Systematic collection	Integrated collection	Automated collection	Continuous collection
	Minimal analysis	Basic analysis	Advanced analysis	Predictive analysis	Adaptive analysis
Cost Knowledge	Traditional cost accounting	Enhanced cost tracking	Episode-based costing	Accurate activity-based costing	Predictive cost modelling
	Limited transparency	Improved transparency	Full transparency	Real-time transparency	Value transparency
	Department-level analysis	Service-line analysis	Condition-level analysis	Patient-level analysis	Personalized analysis
Knowledge Integration	Siloed knowledge	Connected systems	Integrated platforms	Unified architecture	Adaptive architecture
	Limited sharing	Enhanced sharing	Systematic sharing	Pervasive sharing	Frictionless sharing
	Manual integration	Facilitated integration	Automated integration	Intelligent integration	Cognitive integration
Knowledge Application	Limited at point of care	Available at point of care	Embedded in workflow	Context-aware delivery	Personalized delivery
	Provider-focused	Team-focused	Patient-team focused	Patient-centered	Person-directed
	Reactive usage	Responsive usage	Proactive usage	Predictive usage	Adaptive usage

continued on following page

Table 6. Continued

Domain	Level 1: Initial	Level 2: Developing	Level 3: Defined	Level 4: Managed	Level 5: Optimizing
Knowledge Governance	Informal processes	Defined processes	Standardized processes	Optimized processes	Adaptive processes
	Limited oversight	Structured oversight	Systematic oversight	Value-based oversight	Continuous oversight
	Reactive management	Active management	Strategic management	Proactive management	Transformative management

CONCLUSION

This chapter has explored the strategic dimensions of KM particularly in healthcare organizations. Starting with an examination of generic strategic frameworks, the chapter demonstrated how Porter's generic strategies, the RBV, and dynamic capabilities can inform KM approaches. The chapter then explored industry analysis techniques, including Porter's Five Forces and PESTEL analysis, which help organizations understand the external knowledge environment.

Internal analysis was examined through the lens of Porter's Value Chain model, adapted as a knowledge value chain for healthcare organizations. This approach helps organizations identify where and how knowledge adds value across the care delivery process. The Fishbone Analysis and McFarlan's Strategic Grid were introduced as tools for diagnosing KM challenges and opportunities.

The chapter provided a comprehensive framework for designing a KM strategy, outlining a structured process from initial assessment to implementation and evaluation. Different strategic approaches such as top-down, bottom-up, and hybrid were examined, along with knowledge strategy models such as codification, personalization, and balanced.

Finally, the chapter explored the critical role of strategic KM in enabling value-based healthcare transformation. KM requirements for different value-based care models were discussed, along with mechanisms through which KM can drive value-based transformation in terms of redesign, performance improvement, and patient engagement.

As healthcare organizations navigate the complex challenges of delivering high-quality, cost-effective care, strategic KM represents a critical capability for sustainable success. By developing sophisticated approaches to knowledge generation, integration, and application, healthcare organizations can enhance clinical outcomes, improve patient experience, optimize resource utilization, and accelerate innovation.

REFERENCES

Ali, I., & Kannan, D. (2022). Mapping research on healthcare operations and supply chain management: A topic modelling-based literature review. *Annals of Operations Research, 315*(1), 29–55. DOI: 10.1007/s10479-022-04596-5 PMID: 35382453

Bekbolatova, M., Mayer, J., Ong, C. W., & Toma, M. (2024). Transformative potential of AI in healthcare: definitions, applications, and navigating the ethical landscape and public perspectives. H*ealthcare*, Benzaghta, M. A., Elwalda, A., Mousa, M. M., Erkan, I., & Rahman, M. (2021). SWOT analysis applications: An integrative literature review. *Journal of Global Business Insights*, *6*(1), 54–72.

Bobruk, M., Wiśniewski, Z., & Kot, A. (2023). Knowledge Management in Hospitals. Review of Literature. *Review of Medical Practice*, *29*(3), 72–80.

Carayannis, E. G., Ferreira, J. J., & Fernandes, C. (2021). A prospective retrospective: Conceptual mapping of the intellectual structure and research trends of knowledge management over the last 25 years. *Journal of Knowledge Management*, *25*(8), 1977–1999. DOI: 10.1108/JKM-07-2020-0581

Cassia, A. R., Costa, I., da Silva, V. H. C., & de Oliveira Neto, G. C. (2020). Systematic literature review for the development of a conceptual model on the relationship between knowledge sharing, information technology infrastructure and innovative capability. *Technology Analysis and Strategic Management*, *32*(7), 801–821. DOI: 10.1080/09537325.2020.1714026

Colombani, F., Encrenaz, G., Sibe, M., Quintard, B., Ravaud, A., & Saillour-Glenisson, F. (2022). Development of an evidence-based reference framework for care coordination with a focus on the micro level of integrated care: A mixed method design study combining scoping review of reviews and nominal group technique. *Health Policy (Amsterdam)*, *126*(3), 245–261. DOI: 10.1016/j.healthpol.2022.01.003 PMID: 35063324

Corbo, L., Kraus, S., Vlačić, B., Dabić, M., Caputo, A., & Pellegrini, M. M. (2023). Coopetition and innovation: A review and research agenda. *Technovation*, *122*, 102624. DOI: 10.1016/j.technovation.2022.102624

Evans, J. M., Brown, A., & Baker, G. R. (2015). Intellectual capital in the healthcare sector: A systematic review and critique of the literature. *BMC Health Services Research*, *15*(1), 1–14. DOI: 10.1186/s12913-015-1234-0 PMID: 26670123

Faroukhi, A. Z., El Alaoui, I., Gahi, Y., & Amine, A. (2020). Big data monetization throughout Big Data Value Chain: A comprehensive review. *Journal of Big Data*, *7*(1), 1–22. DOI: 10.1186/s40537-019-0281-5

Gellweiler, C. (2018). Cohesion of RBV and industry view for competitive positioning. *Strategic Management-International Journal of Strategic Management and Decision Support Systems in Strategic Management*, 23(2).

Golden, S. H., Hager, D., Gould, L. J., Mathioudakis, N., & Pronovost, P. J. (2017). A gap analysis needs assessment tool to drive a care delivery and research agenda for integration of care and sharing of best practices across a health system. *Joint Commission Journal on Quality and Patient Safety*, *43*(1), 18–28. DOI: 10.1016/j.jcjq.2016.10.004 PMID: 28334581

Hulse, N. C., Galland, J., & Borsato, E. P. (2012). *Evolution in clinical knowledge management strategy at Intermountain Healthcare*. AMIA Annual Symposium Proceedings, Jalundhwala, F., & Londhe, V. (2023). A systematic review on implementing operational excellence as a strategy to ensure regulatory compliance: A roadmap for Indian pharmaceutical industry. *International Journal of Lean Six Sigma*, *14*(4), 730–758.

Kothari, A., Hovanec, N., Hastie, R., & Sibbald, S. (2011). Lessons from the business sector for successful knowledge management in health care: A systematic review. *BMC Health Services Research*, *11*(1), 1–11. DOI: 10.1186/1472-6963-11-173 PMID: 21787403

Mahdi, O. R., Almsafir, M. K., & Yao, L. (2011). The role of knowledge and knowledge management in sustaining competitive advantage within organizations: A review. *African Journal of Business Management*, *5*(23), 9912.

Martinez, V. A., Betts, R. K., Scruth, E. A., Buckley, J. D., Cadiz, V. R., Bertrand, L. D., Paulson, S. S., Dummett, B. A., Abhyankar, S. S., Reyes, V. M., Hatton, J. R., Sulit, R., & Liu, V. X. (2022). The Kaiser Permanente Northern California Advance Alert Monitor Program: An automated early warning system for adults at risk for in-hospital clinical deterioration. *Joint Commission Journal on Quality and Patient Safety*, *48*(8), 370–375. DOI: 10.1016/j.jcjq.2022.05.005 PMID: 35902140

Milstein, N., Striet, Y., Lavidor, M., Anaki, D., & Gordon, I. (2022). Rivalry and performance: A systematic review and meta-analysis. *Organizational Psychology Review*, *12*(3), 332–361. DOI: 10.1177/20413866221082128

Ndzabukelwako, Z., Mereko, O., Sambo, T., & Thango, B. (2024). The Impact of Porter's Five Forces Model on SMEs Performance: A Systematic Review. Available at *SSRN* 4999059.

Nuti, S., & Vainieri, M. (2014). Strategies and tools to manage variation in regional governance systems. Medical Practice Variations. Health Services Research. Boston: Springer.

O'Hara, N. N., Nophale, L. E., O'Hara, L. M., Marra, C. A., & Spiegel, J. M. (2017). Tuberculosis testing for healthcare workers in South Africa: A health service analysis using Porter's Five Forces Framework. *International Journal of Healthcare Management, 10*(1), 49–56. DOI: 10.1080/20479700.2016.1268814

Olsen, G., Quam, J., Wolfe, D., Soria, N., West, M., Gibbons, S., Province, W., Brunisholz, K. D., Belnap, T., Knighton, A. J., Allen, L., Pollard, M., Woodruff, M., & Srivastava, R. (2021). Quality improvement education: Redesigning Intermountain Healthcare's Advanced Training Program for a value-based learning health care system. *Quality Management in Health Care, 30*(3), 209–211. DOI: 10.1097/QMH.0000000000000327 PMID: 34048376

Omotayo, F. O. (2015). Knowledge Management as an important tool in Organisational Management: A Review of Literature. *Library Philosophy and Practice, 1*, 1–23.

Peeters, L. M., van Munster, C. E., Van Wijmeersch, B., Bruyndonckx, R., Lamers, I., Hellings, N., Popescu, V., Thalheim, C., & Feys, P. (2019). Multidisciplinary data infrastructures in multiple sclerosis: Why they are needed and can be done! *Multiple Sclerosis Journal, 25*(4), 500–509. DOI: 10.1177/1352458518807076 PMID: 30381984

Pflugfelder, N. S. (2021). Knowledge management as a driver of performance in ambulatory healthcare–a systematic literature review through an intellectual capital lens. *Journal of Intellectual Capital, 22*(2), 403–432. DOI: 10.1108/JIC-02-2020-0068

Pike, B., & Mongan, D. (2014). *The integration of health and social care services.* https://www.lenus.ie/bitstream/handle/10147/315536/HRB%20The_integration_of_health_and_social_care_services_2014.pdf?sequence=1

Shellum, J. L., Nishimura, R. A., Milliner, D. S., Harper, C. M.Jr, & Noseworthy, J. H. (2017). Knowledge management in the era of digital medicine: A programmatic approach to optimize patient care in an academic medical center. *Learning Health Systems, 1*(2), e10022. DOI: 10.1002/lrh2.10022 PMID: 31245559

Stroetmann, B., & Aisenbrey, A. (2012). Medical knowledge management in healthcare industry. *World Academy of Science, Engineering and Technology, 64*, 557–562.

Zanotto, B. S., da Silva Etges, A. P. B., Marcolino, M. A. Z., & Polanczyk, C. A. (2021). Value-based healthcare initiatives in practice: A systematic review. *Journal of Healthcare Management, 66*(5), 340–365. DOI: 10.1097/JHM-D-20-00283 PMID: 34192716

Chapter 10
Preventative Care

ABSTRACT

Preventive healthcare is critical for reducing chronic disease burden, leveraging digital tools like mobile health (mHealth) applications, wearables, telemedicine, and artificial intelligence (AI)-driven analytics to enhance primary, secondary, tertiary, and quaternary prevention. This chapter explores an integrated approach combining these technologies with knowledge management (KM) strategies—such as big data analytics, clinical decision support systems (CDSS), and predictive models—to deliver personalized, proactive interventions. Case studies, including Sweetch and BetaMe, illustrate efficacy, achieving outcomes like weight loss and glycated hemoglobin (HbA1c) reduction compared to standard care. Advanced techniques, such as machine learning and behavioral intervention technologies (BITs), refine precision prevention by analyzing real-world data for early risk detection and adaptive behavior change.

INTRODUCTION

Prevention is the foundation of healthcare, shifting the focus from reactive treatment to proactive health management (Waldman & Terzic, 2019). According to the World Health Organization (2024), noncommunicable diseases (NCDs) such as cardiovascular diseases, cancers, chronic respiratory diseases, and diabetes accounted for 75% of global non-pandemic-related deaths in 2021, totalling 43 million deaths. Approximately 80% of premature NCD deaths related to these four disease categories, driven by various risk factors such as tobacco use, unhealthy diets, harmful alcohol consumption, insufficient physical activity, and air pollution. Evidence from the American Diabetes Association (2021) suggests that addressing these factors through lifestyle interventions, like healthier eating and sufficient physical activities, can significantly reduce the burden of these diseases and also it highlights the need for structured, technology-driven preventive strategies. Recent years have

DOI: 10.4018/979-8-3373-1877-6.ch010

seen growing interest in digital health solutions to strengthen prevention. Advances in knowledge management (KM) tools along with artificial intelligence (AI), and big data analytics have rewritten the prevention concept, enabling personalized and timely interventions (Park et al., 2019). A scoping review conducted by Willis et al. (2022) examined digital health interventions in preventive care within primary care settings, revealing that these technologies significantly enhance decision-making, promote behavioral change, and improve self-care, particularly for chronic disease prevention. Key technologies included electronic health records (EHRs) (68.9%), clinical decision support (36.5%), and telehealth (36.5%). While digital health interventions demonstrated meaningful improvements in clinical outcomes (70.1%) and statistically significant benefits (79.7%), focusing on alerts for care management.

The purpose of this chapter is to examine how digital health interventions, integrated with KM strategies, support different levels of prevention, namely primary, secondary, tertiary and quaternary prevention. These respective sections of this chapter will be structured as follows. In the first section, the chapter will outline the types of prevention in healthcare. Next, it will explore the role of digital health technologies in enhancing preventive care, followed by an examination of the importance of KM in supporting personalized, precision prevention. Finally, it will present a case example, including diabetes self-care, to illustrate real-world applications of intelligent knowledge-based preventive strategies.

TYPES OF PREVENTION

Prevention in healthcare is traditionally categorized into three distinct levels namely primary, secondary and tertiary prevention (AbdulRaheem, 2023). However, more recent literature has emerged that offers addition of quaternary prevention as a fourth level (Durrer Schutz et al., 2019). Table 1 provides a brief overview of these four levels of prevention with their respective characteristics.

Table 1. Levels of prevention in healthcare

Type of Prevention	Definition	Key Objectives	Examples
Primary Prevention	Aimed at avoiding the manifestation of a disease	Reduce incidence of disease through risk reduction	Vaccinations, smoking cessation programs, healthy eating, decreasing physical inactivity
Secondary Prevention	Early detection of disease when it is asymptomatic	Reduce disease severity and complications through early intervention	Screening programs, regular check-ups, monitoring of risk factors
Tertiary Prevention	Management of established disease to minimize disability	Slow disease progression and prevent complications	Cardiac rehabilitation programs, diabetes self-management education
Quaternary Prevention	Protection from unnecessary medical interventions	Prevent overdiagnosis and overtreatment	Careful use of diagnostic testing, avoidance of unnecessary procedures

Primary Prevention

According to the Wonca International Dictionary for General/Family Practice, primary prevention is defined as, "Action taken to avoid or remove the cause of a health problem in an individual or a population before it arises." Primary prevention represents the most efficient use of healthcare resources, as it eliminates not only the burden of suffering but also the costs associated with treatment (Briggs et al., 2016). This includes activities related to health promotion and specific protection, such as immunization (e.g., polio vaccination) and lifestyle modifications (e.g., healthy diet, exercise, smoking cessation). The goal is to prevent the onset of disease before any symptoms or risk factors develop.

Health promotion activities typically involve educational and behavioral interventions designed to encourage healthy lifestyle choices and create supportive environments for health. Examples include nutritional education programs, physical activity initiatives, and tobacco control. Numerous public health interventions have demonstrated that community-based health promotion programs play a crucial role in reducing modifiable risk factors for chronic NCDs (Effendi et al., 2023; Mundra et al., 2023). Vaccination programs represent an example of specific protection with significant efficacy in preventing infectious diseases (Larson et al., 2022). It is now well established from a variety of studies that immunization programs have dramatically reduced the global burden of diseases such as polio, measles, and diphtheria (Orenstein & Ahmed, 2017).

In recent years, there has been an increasing interest in the application of digital technologies to enhance primary prevention efforts. mHealth applications, wearable devices, and social media platforms have emerged as influential tools in promoting physical activity and facilitating behavior change. Research has demonstrated that these emerging technologies enhance engagement with lifelong physical activity by integrating personalized health promotion strategies. Systematic reviews and empirical studies further highlight their effectiveness in assessing activity levels, promoting exercise, and offering innovative solutions for integrating technology into physical activity interventions. (Gao & Lee, 2019).

Secondary Prevention

Secondary prevention involves the early detection and prompt intervention for diseases that have not yet become symptomatic or have reached early stages of development. The primary objective of secondary prevention is to identify diseases in their nascent stages when treatment is most effective and complications can be averted or minimized (Razzak et al., 2020).

Screening programs allow healthcare providers to detect conditions such as cancer, diabetes, and cardiovascular disease before clinical indicators emerge (Celermajer et al., 2012). Many studies have been published on the efficacy of screening interventions in reducing mortality and morbidity from various conditions. For instance, Weedon-Fekjær et al. (2014) conducted a prospective cohort study to evaluate the effectiveness of contemporary mammography screening in reducing breast cancer mortality. By analyzing data from Norwegian women aged 50-79 years between 1986 and 2009, the study found that invitation to biennial mammography screening was associated with a 28% reduction in breast cancer mortality (mortality rate ratio: 0.72, 95% CI: 0.64 to 0.79). The findings suggest that large-scale screening programs play a significant role in lowering breast cancer deaths. Additionally, Madimenos et al. (2022) have shown that mobile screening units equipped with point-of-care testing capabilities can effectively reach underserved populations in remote or economically disadvantaged areas.

The digital revolution has transformed secondary prevention by enhancing the precision, accessibility, and efficiency of screening programs. Artificial intelligence (AI) algorithms have demonstrated over 90% accuracy in diagnosing diseases from medical images, significantly transforming healthcare by enhancing medical imaging, diagnosis, and patient care. In certain domains, AI has even outperformed human experts in detecting abnormalities, streamlining healthcare management and surgical procedures. As AI continues to grow, its continuous collaboration with healthcare professionals lead to superior patient outcomes and a revolutionized approach to medical diagnostics (Eskandar, 2023).

An important development in secondary prevention is the emergence of risk assessment models that integrate multiple data points to identify individuals at elevated risk of developing specific conditions. These models typically incorporate demographic information, family history, genetic markers, and physiological parameters to generate personalized risk profiles (Weitzel et al., 2011). KM systems that facilitate the collection, analysis, and interpretation of these diverse data elements play a crucial role in operationalizing risk-based screening approaches.

Tertiary Prevention

Tertiary prevention focuses on managing established diseases to minimize complications, prevent disability, and restore function. This level of prevention is particularly relevant for chronic conditions such as diabetes, congestive heart failure, and chronic obstructive pulmonary disease, which require ongoing management to optimize outcomes and quality of life (de Bruin et al., 2011).

In the context of tertiary prevention interventions, rehabilitation programs, self-management education, and medication adherence initiatives being widely accepted components. Cardiac rehabilitation programs, for instance, Taylor et al. (2022) have demonstrated that cardiac rehabilitation is a clinically and cost-effective intervention for patients with heart disease, significantly reducing recurrent cardiovascular events and improving functional capacity, well-being, and health-related quality of life. Global access to cardiac rehabilitation remains limited, further worsened by the COVID-19 pandemic. Emerging evidence supports home-based and technology-driven models as viable alternatives or adjuncts to traditional center-based programs, particularly in low- and middle-income countries where scalable and affordable solutions are urgently needed.

The digital transformation of tertiary prevention has been characterized with the remote monitoring technologies, telehealth platforms, and digital therapeutic applications designed to support disease management beyond traditional healthcare settings. Comparing traditional and technology-enhanced tertiary prevention, research has found that digital health interventions, particularly mobile applications and computer-based solutions, can significantly improve clinical outcomes, enhance patient engagement, and reduce healthcare utilization across various chronic disease contexts. A systematic review of digital health studies revealed that these interventions are particularly effective in improving health behaviours, clinical assessment, treatment compliance, and care coordination (Ibrahim et al., 2022).

KM plays a critical role in tertiary prevention by enabling the systematic collection, analysis, and dissemination of patient-generated health data and clinical observations. Effective KM systems in the digital transformation of healthcare by enabling healthcare providers to detect patient status deteriorations, identify patterns linked to poor outcomes, and implement timely interventions to enhance care delivery and prevent adverse events (Alhashmi, 2025).

Quaternary Prevention

Quaternary prevention focuses on protecting individuals from excessive or unnecessary medical interventions (Martins et al., 2018) and it is the action taken to identify patients at risk of overmedicalisation, to protect them from new medical invasion, and to suggest to them interventions which are ethically acceptable (Tesser & Norman, 2021). The principal objective of quaternary prevention is to mitigate the potential harms associated with medical interventions that offer minimal or no benefit relative to their risks and costs. The importance of quaternary prevention has been emphasized by growing evidence of medical overuse across various healthcare domains. For instance, Sartelli et al. (2020) highlighted that antimicrobial resistance (AMR) is exacerbated by inappropriate antibiotic use and poor infection control strategies, particularly in low and middle-income countries (LMICs), where differences in healthcare systems and prescribing practices may contribute to unnecessary or inappropriate antibiotic prescriptions. KM tools and techniques are essential for implementing quaternary prevention by facilitating evidence-based decision-making and promoting appropriate use criteria.

ROLE OF DIGITAL HEALTH SOLUTIONS IN SUPPORTING PREVENTION

In recent years, there has been an increasing interest in the application of digital technologies to enhance preventive healthcare delivery. The digital transformation of prevention has been driven by advances in data analytics, AI, mobile computing, and internet connectivity, which collectively enable more personalized, accessible, and efficient preventive interventions (Neto & Von Wyl, 2024).

Figure 1 illustrates the how the digital health solutions that support prevention across its various levels. This system involves technologies ranging from consumer facing applications to sophisticated clinical decision support system (CDSS), all interconnected through data exchange networks and KM infrastructures.

Figure 1. Digital health solutions for prevention

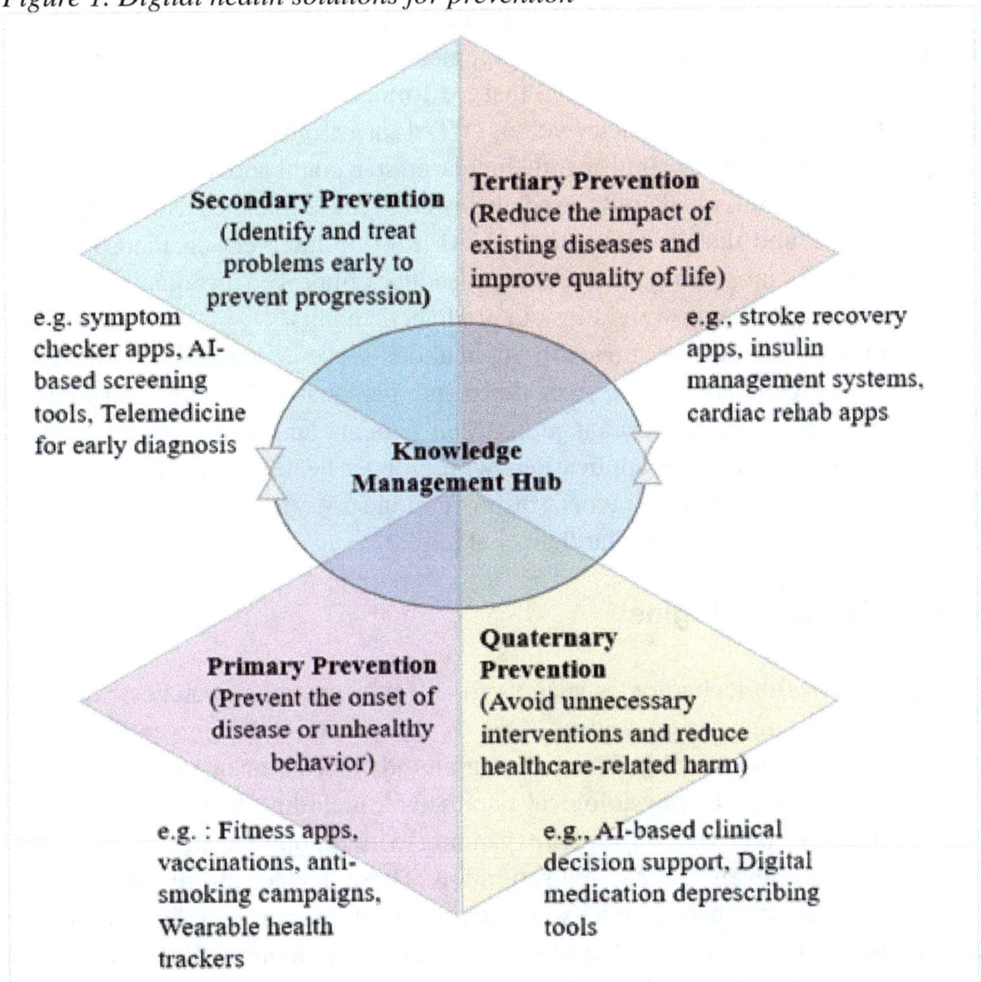

mHealth Applications

mHealth applications represent one of the most popular digital health technologies in preventive care. With the increased use of smartphones worldwide, these applications offer unique opportunities to deliver personalized prevention interventions.

Recent research has demonstrated that mHealth applications can effectively support various preventive behaviours, including physical activity, healthy eating, smoking cessation, and medication adherence. However, despite the availability of evidence-based mHealth apps, user adherence remains a significant challenge, with an average adherence rate of 56.0% (SD 24.4%) (Jakob et al., 2022). The effectiveness of mHealth interventions for prevention has been recognised to several features, including real-time feedback, behavioral nudging, social support facilitation, and gamification elements that enhance engagement (Orte et al., 2023). For example, Vetrovsky et al. (2023) demonstrated that a 12-month mHealth intervention integrating just-in-time adaptive interventions (JITAI) principles, Fitbit-based tracking, and personalized feedback significantly increased step count and reduced sedentary behavior among individuals with prediabetes and type 2 diabetes. Similarly, Han et al. (2024) found that an mHealth-enabled lifestyle intervention incorporating coaching, dietary monitoring, and real-time feedback improved diet quality and was positively correlated with weight loss in prediabetic adults.

From a KM perspective, mHealth applications serve as valuable conduits for collecting user-generated health data, delivering personalized health information, and facilitating knowledge exchange between patients and healthcare providers. The integration of mHealth applications with broader health information systems allows for the continuous refinement of prevention strategies based on real-world usage patterns and outcomes (Bardhan et al., 2020).

Wearable Technologies

Wearable health technologies such as fitness trackers, smartwatches, and specialized medical monitoring devices serve as valuable tools for preventive health monitoring. Wearable health devices integrate advanced sensor technologies to capture a wide range of physiological parameters, including heart rate, physical activity, and sleep patterns, while also expanding to more sophisticated metrics such as blood glucose levels and oxygen saturation. These devices are revolutionizing health monitoring by enabling both self-tracking for fitness and providing clinicians with real-time data for earlier diagnosis and treatment guidance. With continuous advancements wearable health devices are playing a critical role in transforming healthcare, particularly in cardiovascular monitoring and the development of smart medical wearables like sensor-equipped garments (Dias & Paulo Silva Cunha, 2018).

The preventive value of wearable health devices lies in their ability to provide continuous health monitoring outside clinical settings, enabling early detection of physiological changes and facilitating timely interventions. Additionally, real-time feedback from wearables enhances user engagement, promotes adherence to preventive behaviours, and complements digital health interventions like the PRIME program, which integrates mobile applications for self-monitoring and education in pre-diabetes management (Teoh et al., 2024). Studies have demonstrated that the use of sensor-based remote patient monitoring (RPM) and wearable health devices has led to significant improvements in physical activity, glycemic control, and engagement with digital health programs (Everett et al., 2018; Graue et al., 2024).

In the prevention of diabetes complications such as diabetic foot ulcers, RPM has been shown to enhance early detection and intervention, reducing the risk of severe outcomes. Integrating RPM into comprehensive disease management strategies allows for proactive monitoring, contextualized clinical assessments, and predictive system management, ultimately supporting better patient outcomes and optimizing diabetes care (Everett et al., 2018; Graue et al., 2024).

While wearables generate valuable real-time health data for personalized prevention strategies, challenges remain in technology access, digital literacy, and usability, particularly among older adults and underserved populations (Lawrence et al., 2021; Summers et al., 2021). Effective deployment requires robust systems for data integration, validation, and actionability within clinical workflows (Nguyen et al., 2024).

Telemedicine and Virtual Care

Telemedicine and virtual care platforms enhance access to preventive services by overcoming geographical barriers. Their adoption accelerated during the COVID-19 pandemic, transforming preventive care delivery (Del Rocio Sevilla-Gonzalez et al., 2022).

Telehealth modalities include video consultations, remote monitoring, store-and-forward technologies, and mHealth applications with clinical oversight (Castro Sweet et al., 2018). These approaches have been applied to diabetes prevention programs, lifestyle counselling, medication adherence, and chronic disease monitoring. Randomized controlled trials have shown that mHealth-enabled telemedicine interventions improve weight loss, physical activity, and glycemic control outcomes (Teoh et al., 2024).

KM systems facilitate secure health information exchange, integrate remote monitoring data with EHRs, and support clinical decision-making in virtual settings (Lawrence et al., 2021; Nguyen et al., 2024). However, digital literacy, internet accessibility, and healthcare provider engagement remain barriers to adoption (Chung et al., 2023; Nurul Fatihah Mohd et al., 2023). Culturally tailored interventions and improved technological usability are critical for maximizing impact.

AI and Predictive Analytics

AI and predictive analytics transform preventive healthcare by processing heterogeneous data, identifying disease risk patterns, and generating personalized prevention recommendations with unprecedented precision (Ackermann et al., 2014; Haider et al., 2019).

Applications span risk prediction, screening optimization, behavior change support, and resource allocation. Everett et al. (2018) demonstrated that AI-based recommendations with real-time weight monitoring improved engagement among individuals with prediabetes. Toro-Ramos et al. (2020) showed mobile-based digital diabetes prevention programs leveraging AI-driven coaching were effective in promoting weight loss. Predictive models using AI have been deployed for diabetes, cardiovascular diseases, and metabolic disorders (Nurul Fatihah Mohd et al., 2023). These models integrate EHRs, wearable data, environmental factors, and lifestyle behaviors to generate individualized risk assessments. AI-enabled dietary monitoring and real-time feedback have effectively improved diet quality and glycemic control (Lim et al., 2022).

From a KM perspective, AI systems both store and generate preventive health insights by synthesizing patient-generated data, operationalizing clinical expertise into algorithmic decision-making, and continuously refining models based on outcomes (Kim et al., 2019; Novak et al., 2024). Effective implementation requires frameworks for algorithm validation, transparency, and responsible deployment. Challenges include algorithmic bias, data privacy concerns, and the need for explainable AI models (Abbaspoor et al., 2020; Haider et al., 2019).

IMPORTANCE OF KM FOR SUPPORTING PERSONALIZED PRECISION PREVENTION

KM plays a crucial role in translating the potential of digital health technologies into meaningful improvements in personalized precision prevention by ensuring that the right knowledge is available at the right time for informed decision-making. Effective KM implementation in healthcare leverages tools such as health information and communication technology, CDSS, EHRs, and communities of practice to enhance knowledge sharing and utilization. By integrating real-time data, predictive analytics, and patient-specific insights, KM enables healthcare systems to deliver targeted, proactive, and individualized preventive interventions. This approach not only improves patient outcomes and risk mitigation but also helps overcome barriers to knowledge accessibility, ultimately fostering a more efficient, data-driven, and patient-centric healthcare environment (Alam et al., 2024; Shahmoradi et al., 2017).

Knowledge Creation and Capture

The first step in leveraging Knowledge Management (KM) for preventive healthcare is the systematic creation and capture of relevant knowledge from diverse sources. These sources include scientific literature, clinical guidelines, electronic health records (EHRs), patient-generated health data, and real-world evidence from implemented interventions (Fernandez-Luque et al., 2022). Effectively capturing this knowledge enables healthcare systems to develop evidence-based, data-driven preventive strategies that address emerging health risks.

Knowledge creation in preventive care is increasingly characterized by collaborative and interdisciplinary methods that integrate insights from medicine, public health, behavioral science, data analytics, and implementation research. This integrated approach enables a more comprehensive understanding of disease mechanisms, risk factors, intervention effectiveness, and implementation considerations (Bedson et al., 2021).

Digital technologies have revolutionized knowledge creation by enabling the collection and analysis of previously inaccessible or underutilized data sources. For instance, social media analytics have been leveraged to track disease outbreaks, monitor health behaviors, and assess public sentiment toward preventive interventions such as vaccinations (Fernandez-Luque et al., 2022). These advancements highlight the importance of KM in harnessing real-time data to inform timely and proactive public health responses.

Table 2 summarizes the key knowledge sources that inform personalized prevention strategies and the digital technologies that facilitate their capture and integration.

Table 2. Knowledge sources for personalized prevention

Knowledge Source	Description	Enabling Technologies	Value for Prevention
EHRs	Clinical documentation, test results, medications, procedures	EHR systems, Natural Language Processing	Identification of risk factors, tracking preventive service delivery, clinical decision support
Patient-Generated Health Data	Self-reported symptoms, behaviors, biometric measurements	Mobile apps, wearables, patient portals	Real-time monitoring, behavioral insights, engagement with prevention
Environmental Data	Air quality, weather patterns, built environment characteristics	Internet of things (IoT) sensors, satellite imagery	Contextual risk assessment, environmental intervention design
Social Determinants Data	Socioeconomic status, education, social support, community resources	Social needs screening tools, public datasets, community mapping	Addressing upstream factors, targeting interventions to vulnerable populations

continued on following page

Table 2. Continued

Knowledge Source	Description	Enabling Technologies	Value for Prevention
Scientific Literature	Research findings, clinical guidelines, systematic reviews	Literature databases, automatic summarization tools	Evidence-based intervention design, best practice dissemination
Real-World Evidence	Outcomes data from implemented programs, practice-based observations	Health information exchanges, registries, learning health systems	Continuous improvement, comparative effectiveness, implementation insights

Knowledge Organization and Storage

Once relevant preventive health knowledge is captured, it must be systematically organized and stored in accessible repositories to facilitate efficient retrieval and application. Knowledge organization involves classification, indexing, tagging, and structuring information to enhance its findability and usability in preventive care (Zeng et al., 2020).

In preventive healthcare, effective knowledge organization strategies rely on standardized terminologies, ontologies, and data models to integrate heterogeneous information sources (Shahzad et al., 2021). Ontologies enable semantic relationships between different data points, supporting interoperability across smart healthcare services, digital health platforms, and clinical decision-making tools.

The adoption of cloud-based knowledge repositories has further transformed preventive health knowledge storage, offering scalable, flexible, and secure platforms that ensure seamless data sharing across organizational boundaries. These repositories integrate structured and unstructured data, support version control, and implement role-based access mechanisms to align with privacy regulations and institutional policies (Shahzad et al., 2021) By leveraging advanced knowledge organization systems, such as taxonomies, classification schemes, and ontologies, healthcare systems can improve information retrieval, enhance collaboration, and support evidence-based preventive health strategies.

Knowledge Dissemination and Application

The ultimate value of preventive health knowledge lies in its timely dissemination to relevant stakeholders and effective application in clinical and self-care contexts.

Knowledge dissemination strategies in preventive care include CDSS, patient education platforms, provider training programs, and policy guidance (Krist et al., 2017).

CDSS represent a vital source of knowledge application in preventive care. These systems integrate evidence-based guidelines, patient-specific data, and algorithmic logic to generate personalized prevention recommendations at the point of care (Sharma & Kaushik, 2025). For example, Buzancic et al. (2024) found that CDSS implementations significantly improved adherence to preventive service guidelines for vaccinations, cancer screening, and cardiovascular risk management.

Patient-facing knowledge dissemination channels have expanded dramatically with the advent of digital health technologies. Patient portals, mobile applications, conversational agents, and multimedia educational platforms now enable more personalized, engaging, and accessible delivery of preventive health information (Alzughaibi, 2023). These digital channels allow for tailoring of content based on individual risk profiles, preferences, literacy levels, and readiness for behavior change.

Knowledge Evaluation and Refinement

The dynamic nature of preventive health knowledge necessitates continuous evaluation and refinement of knowledge assets based on emerging evidence, real-world outcomes, and evolving contextual factors. Knowledge evaluation processes should assess the accuracy, relevance, applicability, and impact of preventive knowledge resources in diverse healthcare contexts. Learning health systems represent an advanced approach to knowledge evaluation and refinement, creating virtuous cycles where routine care generates data that informs continuous improvement of preventive strategies (Liu et al., 2016). These systems typically incorporate performance metrics, patient-reported outcomes, implementation data, and comparative effectiveness analyses to identify opportunities for optimizing prevention approaches.

Figure 2 illustrates the KM cycle for personalized precision prevention, highlighting the iterative and interconnected nature of knowledge processes in preventive care.

Figure 2. KM cycle for personalized precision prevention

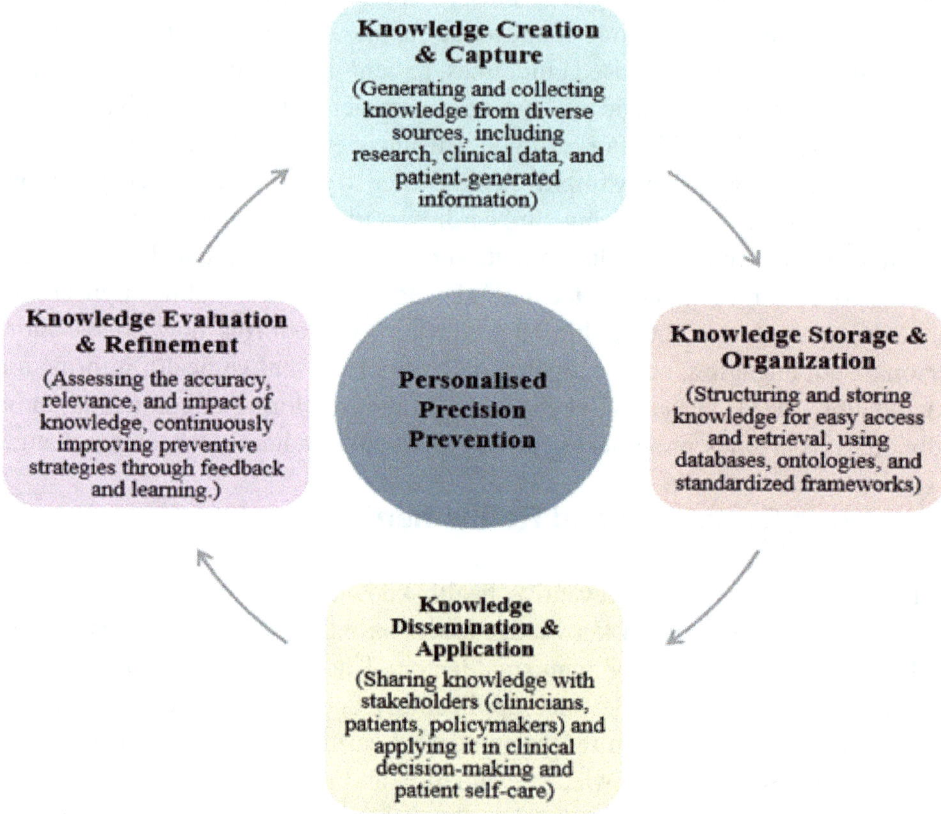

TECHNIQUES AND TECHNOLOGIES FOR SUPPORTING PERSONALIZED PRECISION PREVENTION

Personalized precision prevention represents the convergence of several technological capabilities and methodological approaches that together enable more targeted, individualized preventive interventions. Unlike traditional prevention strategies, precision prevention aims to tailor preventive measures to individuals based on their unique risk profiles, preferences, and contexts (de Toro-Martín et al., 2017). This section examines the key techniques and technologies that support the implementation of personalized precision prevention in contemporary healthcare settings.

Advanced Analytics and Machine Learning (ML)

Advanced analytics and ML techniques provide the computational foundation for translating complex, multimodal data into actionable prevention insights (Ahmed et al., 2023). These techniques enable pattern recognition across diverse data types, risk stratification based on multiple factors, and predictive modeling of future health trajectories that inform preventive interventions.

Several categories of advanced analytics are particularly relevant for precision prevention as listed below.

- Predictive Analytics: Statistical and ML methods that identify patterns in historical data to forecast future health risks and outcomes, enabling proactive preventive intervention (Razzak et al., 2020).
- Prescriptive Analytics: Advanced analytical approaches that not only predict outcomes but also recommend specific courses of action to achieve desired prevention goals (Khandakar et al., 2024).
- Network Analytics: Techniques that analyze relationships between diverse factors such as biological, behavioral, social, and environmental to identify intervention points with maximal preventive impact (Razzak et al., 2020).
- Natural Language Processing (NLP): Computational methods that extract meaningful information from unstructured text data, enabling the incorporation of clinical notes, patient-reported outcomes, and scientific literature into prevention models (Gonzalez-Hernandez et al., 2017).

The application of machine learning in disease prevention has yielded promising outcomes. A study by Cho et al. (2021) found that machine learning-based cardiovascular risk prediction models outperformed traditional statistical approaches. The neural network model achieved a C-statistic of 0.751, surpassing the Pooled Cohort Equation (0.738) and demonstrating improved calibration (Hosmer–Lemeshow $\chi^2 = 86.1$, $P < 0.001$). These findings highlight the potential of machine learning algorithms to enhance accuracy in cardiovascular disease risk assessment and clinical decision-making.

Similarly, Swinckels et al. (2024) demonstrated that deep learning models applied to longitudinal electronic health record (EHR) data could effectively predict the onset of various diseases, particularly type 2 diabetes, kidney diseases, and circulatory disorders, well before clinical diagnosis. Their scoping review of 20 studies published between 2018 and 2022 found that recurrent neural networks and long short-term memory techniques achieved high diagnostic accuracy, enabling earlier detection and intervention. These models leveraged demographic data, symptoms, lab results, medications, and BMI to identify high-risk individuals, offering potential benefits such as personalized healthcare, reduced clinical workload, and improved preventive strategies.

Behavioral Intervention Technologies (BITs)

BITs are digital tools designed to support behavior change and improve health outcomes by making preventive care more engaging and accessible. These technologies help individuals adopt healthier lifestyles, manage chronic conditions, and reduce disease risks through personalized, technology-driven interventions (Schueller et al., 2016).

One key category of BITs is Digital Therapeutics (DTx), which includes software-based interventions such as mobile applications, virtual reality, and wearable devices. These tools offer non-invasive and individualized treatment options for conditions like Alzheimer's disease (AD), where digital solutions such as "Memory Matters", "AlzSense", and "Immersive Virtual Reality (iVR)" help improve cognition, memory, and overall brain health (Manchanda et al., 2022).

Another important category is Adaptive Interventions, which adjust their content, intensity, or delivery method based on a user's progress and needs. These interventions use Sequential Multiple Assignment Randomized Trial (SMART) designs to personalize support and enhance long-term engagement (Nahum-Shani & Almirall, 2019).

Gamified Prevention Applications incorporate game-like elements such as points, leaderboards, and challenges to motivate users to maintain healthy habits. Apps like "bant" (for diabetes management), "mySugr" (which uses health challenges), and "Fitocracy" (which integrates achievement badges) have been shown to increase engagement and adherence to self-care routines (Miller et al., 2016).

Finally, AI-powered conversational agents and chatbots are increasingly being used to provide personalized health recommendations and behavioral coaching. Research has shown that anthropomorphic and avatar-based chatbots can improve user engagement and compliance, though challenges related to data privacy and ethical considerations persist (Anisha et al., 2024).

Overall, BITs play a crucial role in preventive healthcare by offering personalized, engaging, and scalable solutions that help individuals adopt and maintain healthier behaviors. These technologies continue to evolve, integrating AI, gamification, and real-time data tracking to enhance their effectiveness in precision prevention and long-term health management.

Integrated Prevention Platforms

Integrated prevention platforms leverage hardware and software solutions to create scalable, data-driven ecosystems that support personalized and precision prevention strategies. These platforms combine real-time health monitoring, artificial intelligence (AI)-driven analytics, and secure data integration to facilitate risk assessment, intervention delivery, and outcome tracking across various healthcare settings (Hsiao et al., 2022).

A key example of such an approach is a comprehensive healthcare platform for cardiovascular disease prevention, which integrates biomedical data, wearable sensors, and deep-learning algorithms for continuous health monitoring and risk detection. This system enables remote patient supervision by doctors, hospital procedures, and self-care monitoring via personal devices such as smartphones and tablets. Additionally, AI-powered analytics enhance early detection of cardiovascular risks, allowing for personalized intervention strategies and epidemiological profiling (Gómez-García et al., 2021).

The success of integrated prevention platforms depends on their ability to seamlessly connect multiple healthcare touchpoints, ensuring that preventive measures are both scalable and accessible. By incorporating secure data exchange, AI-driven insights, and extended device integration, these platforms enhance healthcare efficiency, improve patient outcomes, and support evidence-based decision-making. As digital health technology advances, integrated platforms will play a pivotal role in delivering proactive, personalized, and precision-driven preventive care solutions.

Implementation Considerations

While the technical capabilities for personalized precision prevention continue to advance rapidly, successful implementation requires careful attention to various non-technical factors that influence adoption, effectiveness, and sustainability in real-world settings (Kronlid et al., 2024). These implementation considerations span ethical, legal, organizational, workforce, and equity dimensions that collectively determine the real-world impact of precision prevention technologies. By leveraging these advanced techniques and technologies, personalized precision prevention has the potential to revolutionize healthcare, enabling more proactive and effective interventions tailored to individual needs.

CASE EXAMPLES

To illustrate the practical application of intelligent knowledge-based approaches in preventive healthcare, this section presents two case examples that demonstrate the integration of digital technologies and KM principles in addressing significant public health challenges.

Digital Health Intervention for Prediabetes Management: The Sweetch App

Prediabetes affects millions of individuals worldwide, serving as a critical precursor to type 2 diabetes if left unaddressed. Digital health interventions leveraging AI-driven KM can play a pivotal role in supporting behavior change and preventing disease progression. One such example is the Sweetch app, evaluated for its feasibility, acceptability, and effectiveness in individuals with prediabetes (Everett et al., 2018). This AI-powered mobile application integrates behavioral analytics, real-time monitoring, and adaptive feedback mechanisms to promote sustained lifestyle modifications.

The KM framework of Sweetch incorporates:

- Knowledge capture: Continuous collection of user data, including physical activity levels, dietary habits, and weight fluctuations, facilitated by a connected digital scale.
- Knowledge processing: AI-driven analytics assess user behavior trends and dynamically adjust recommendations based on adherence patterns and engagement levels.
- Knowledge dissemination: The system delivers tailored lifestyle modification suggestions through personalized notifications and interactive content, encouraging sustained engagement.
- Knowledge application: Users apply the recommendations to adjust their physical activity routines, dietary intake, and weight management strategies.
- Knowledge refinement: Continuous learning from user interactions allows the system to refine and optimize its guidance for long-term efficacy.

The Sweetch intervention demonstrated high user acceptability (86% retention rate) and led to mean weight loss of 1.6 kg ($P<.001$), increased physical activity by 2.8 MET-hours per week ($P=.02$), and a median hemoglobin (HbA1c) reduction of 0.1% (IQR -0.2% to 0.1%; $P=.04$). However, challenges such as low participation in additional lifestyle counselling (only 4%) indicate the need for integrated support mechanisms to enhance adherence and engagement.

This case exemplifies how AI-driven knowledge systems can effectively transform prediabetes management by delivering personalized, adaptive support that empowers individuals to sustain healthy lifestyle changes.

Population-Level Diabetes Prevention: The BetaMe Program

Scaling preventive healthcare interventions across diverse populations requires robust knowledge-driven strategies that integrate digital health tools with primary care workflows. The BetaMe program, a digital intervention evaluated in New Zealand, serves as an example of how knowledge-based frameworks can optimize prediabetes management at the population level (Sarfati et al., 2018).

The BetaMe program's KM infrastructure includes:

- Unified data repository: Integration of digital health records, patient-generated health data, and behavioral tracking logs.
- Evidence engine: Incorporation of culturally tailored educational modules and structured health coaching sessions.
- Prediction factory: Analytics-driven identification of patients at high risk of diabetes progression, guiding personalized intervention strategies.
- Outreach coordination system: Multi-channel communication pathways (SMS, mobile apps, web portals) ensuring timely delivery of recommendations and health coaching.
- Learning framework: Continuous monitoring of engagement metrics, intervention efficacy, and user feedback to refine service delivery.

The program targeted high-risk populations, including Māori and Pacific communities, through a culturally adapted interface, incorporating personalized coaching, goal tracking, and peer support mechanisms. Notable outcomes included 92% completion rate, 91% of participants reducing HbA1c, 94% achieving weight loss (mean: 4.2 kg), and 78% of participants reaching HbA1c levels below the prediabetes threshold. However, barriers such as limited internet access and language differences highlighted the importance of ensuring digital inclusivity in large-scale interventions.

This case highlights the potential of integrated KM in bridging digital health interventions with primary care services, thereby enhancing scalability and accessibility of diabetes prevention efforts across diverse populations.

Diabetes Self-Care

Diabetes self-care is enhanced through digital tools that empower patients to manage their condition proactively. The Sweetch app, previously discussed, supports lifestyle adjustments by delivering AI-driven, personalized recommendations based on real-time activity and dietary data, achieving a median HbA1c reduction of 0.1% (Everett et al., 2018). Similarly, mySugr, a gamified mobile application, improves adherence to glucose monitoring through challenges and rewards, fostering engagement in self-care routines (Miller et al., 2016). These tools leverage KM by capturing patient-generated data (e.g., glucose levels, behaviors), processing it via analytics, and disseminating tailored guidance through user-friendly interfaces. This empowers patients with actionable insights, reducing reliance on clinical oversight. Research shows such interventions improve glycemic control and self-efficacy, though challenges like digital literacy and data privacy require attention (Ibrahim et al., 2022) . These examples highlight how digital health and KM transform diabetes self-care into a proactive, patient-centered process.

CONCLUSION

This chapter explores how digital health solutions and KM revolutionize preventive healthcare across primary, secondary, tertiary, and quaternary levels, shifting focus from disease treatment to proactive management. Technologies such as mHealth applications, wearables, telemedicine, and AI-driven analytics enhance disease prevention, early detection, and management by delivering personalized, accessible interventions. KM underpins precision prevention through tools like big data analytics, CDSS, and predictive models, systematically capturing, organizing, and disseminating health knowledge to optimize decision-making. Case studies, such as Sweetch and BetaMe, demonstrate practical success, with outcomes including significant weight loss and HbA1c reduction. Advanced techniques, including machine learning and behavioral intervention technologies (BITs), further refine prevention by analyzing real-world data for early risk identification and adaptive behavior change. Despite these advances, challenges like privacy risks, AI bias, digital literacy gaps, and data integration barriers persist, necessitating continued research and regulatory frameworks. Together, digital health and KM emerge as powerful drivers of personalized, proactive care, poised to transform precision prevention and improve population health outcomes.

REFERENCES

Abbaspoor, Z., Amani, A., Afshari, P., & Jafarirad, S. (2020). The effect of education through mobile phone short message service on promoting self-care in pre-diabetic pregnant women: A randomized controlled trial. *Journal of Telemedicine and Telecare, 26*(4), 200–206. https://doi.org/https://dx.doi.org/10.1177/1357633X18791419. DOI: 10.1177/1357633X18791419 PMID: 30193565

AbdulRaheem, Y.AbdulRaheem. (2023). Unveiling the significance and challenges of integrating prevention levels in healthcare practice. *Journal of Primary Care & Community Health, 14*, 21501319231186500. DOI: 10.1177/21501319231186500 PMID: 37449436

Ackermann, R. T., Sandy, L. G., Beauregard, T., Coblitz, M., Norton, K. L., & Vojta, D. (2014). A randomized comparative effectiveness trial of using cable television to deliver diabetes prevention programming. *Obesity* (Silver Spring, Md.), 22(7), 1601-1607. https://doi.org/https://dx.doi.org/10.1002/oby.20762

Ahmed, A., Xi, R., Hou, M., Shah, S. A., & Hameed, S. (2023). Harnessing big data analytics for healthcare: A comprehensive review of frameworks, implications, applications, and impacts. *IEEE Access : Practical Innovations, Open Solutions, 11*, 112891–112928. DOI: 10.1109/ACCESS.2023.3323574

Alam, M. A., Sohel, A., Uddin, M., & Siddiki, A. (2024). Big Data and Chronic Disease Management Through Patient Monitoring And Treatment With Data Analytics. *Academic Journal on Artificial Intelligence, Machine Learning. Data Science and Management Information Systems, 1*(01), 77–94.

Alhashmi, S. M. (2025). Knowledge management, health data, and advanced data analytics to expedite solutions in the healthcare industry. In *Digital Healthcare, Digital Transformation and Citizen Empowerment in Asia-Pacific and Europe for a Healthier Society* (pp. 231–247). Elsevier. DOI: 10.1016/B978-0-443-30168-1.00005-0

Alzughaibi, S. (2023). *Utilizing Chatbot Technology to Disseminate Health Information: The Case of Human Papillomavirus Vaccine-User-Centered Design Approach. Wisconsin.edu.* http://digital.library.wisc.edu/1793/93267

American Diabetes Association. (2021). Standards of Medical Care in Diabetes—2021. *Diabetes Care, 44*(Supplement 1), S1–S232. DOI: 10.2337/dc21-Sint PMID: 33298409

Anisha, S. A., Sen, A., & Bain, C. (2024). Evaluating the potential and pitfalls of AI-powered conversational agents as humanlike virtual health carers in the remote management of noncommunicable diseases: Scoping review. *Journal of Medical Internet Research*, *26*, e56114. DOI: 10.2196/56114 PMID: 39012688

Bardhan, I., Chen, H., & Karahanna, E. (2020). Connecting systems, data, and people: A multidisciplinary research roadmap for chronic disease management. *Management Information Systems Quarterly*, *44*(1), 185–200. DOI: 10.25300/MISQ/2020/14644

Bedson, J., Skrip, L. A., Pedi, D., Abramowitz, S., Carter, S., Jalloh, M. F., Funk, S., Gobat, N., Giles-Vernick, T., Chowell, G., de Almeida, J. R., Elessawi, R., Scarpino, S. V., Hammond, R. A., Briand, S., Epstein, J. M., Hébert-Dufresne, L., & Althouse, B. M. (2021). A review and agenda for integrated disease models including social and behavioural factors. *Nature Human Behaviour*, *5*(7), 834–846. DOI: 10.1038/s41562-021-01136-2 PMID: 34183799

Briggs, A. M., Cross, M. J., Hoy, D. G., Sànchez-Riera, L., Blyth, F. M., Woolf, A. D., & March, L. (2016). Musculoskeletal health conditions represent a global threat to healthy aging: A report for the 2015 World Health Organization world report on ageing and health. *The Gerontologist*, *56*(suppl_2), S243–S255. DOI: 10.1093/geront/gnw002 PMID: 26994264

Buzancic, I., Koh, H. J. W., Trin, C., Nash, C., Hadziabdic, M. O., Belec, D., Zoungas, S., Zomer, E., Dalli, L., & Ademi, Z. (2024). Do clinical decision support tools improve quality of care outcomes in the primary prevention of cardiovascular disease: A systematic review and meta-analysis. *American Journal of Preventive Cardiology*, *20*, 100855. DOI: 10.1016/j.ajpc.2024.100855 PMID: 39416379

Castro Sweet, C. M., Chiguluri, V., Gumpina, R., Abbott, P., Madero, E. N., Payne, M., Happe, L., Matanich, R., Renda, A., & Prewitt, T. (2018). Outcomes of a Digital Health Program With Human Coaching for Diabetes Risk Reduction in a Medicare Population. *Journal of Aging and Health*, *30*(5), 692–710. DOI: 10.1177/0898264316688791 PMID: 28553807

Celermajer, D. S., Chow, C. K., Marijon, E., Anstey, N. M., & Woo, K. S. (2012). Cardiovascular disease in the developing world: Prevalences, patterns, and the potential of early disease detection. *Journal of the American College of Cardiology*, *60*(14), 1207–1216. DOI: 10.1016/j.jacc.2012.03.074 PMID: 22858388

Cho, S.-Y., Kim, S.-H., Kang, S.-H., Lee, K. J., Choi, D., Kang, S., Park, S. J., Kim, T., Yoon, C.-H., Youn, T.-J., & Chae, I.-H. (2021). Pre-existing and machine learning-based models for cardiovascular risk prediction. *Scientific Reports*, *11*(1), 8886. DOI: 10.1038/s41598-021-88257-w PMID: 33903629

Chung, H. W., Tai, C. J., Chang, P., Su, W. L., & Chien, L. Y. (2023). The Effectiveness of a Traditional Chinese Medicine–Based Mobile Health App for Individuals With Prediabetes: Randomized Controlled Trial [Article]. *JMIR mHealth and uHealth*, *11*, e41099. Advance online publication. DOI: 10.2196/41099 PMID: 37338977

de Bruin, S. R., Heijink, R., Lemmens, L. C., Struijs, J. N., & Baan, C. A. (2011). Impact of disease management programs on healthcare expenditures for patients with diabetes, depression, heart failure or chronic obstructive pulmonary disease: A systematic review of the literature. *Health Policy (Amsterdam)*, *101*(2), 105–121. DOI: 10.1016/j.healthpol.2011.03.006 PMID: 21592607

de Toro-Martín, J., Arsenault, B. J., Després, J.-P., & Vohl, M.-C. (2017). Precision nutrition: A review of personalized nutritional approaches for the prevention and management of metabolic syndrome. *Nutrients*, *9*(8), 913. DOI: 10.3390/nu9080913 PMID: 28829397

Del Rocio Sevilla-Gonzalez, M., Bourguet-Ramirez, B., Lazaro-Carrera, L. S., Martagon-Rosado, A. J., Gomez-Velasco, D. V., & Viveros-Ruiz, T. L. (2022). Evaluation of a Web Platform to Record Lifestyle Habits in Subjects at Risk of Developing Type 2 Diabetes in a Middle-Income Population: Prospective Interventional Study [Article]. *JMIR Diabetes*, *7*(1), e25105. Advance online publication. DOI: 10.2196/25105 PMID: 35037888

Dias, D., & Paulo Silva Cunha, J. (2018). Wearable health devices—Vital sign monitoring, systems and technologies. *Sensors (Basel)*, *18*(8), 2414. DOI: 10.3390/s18082414 PMID: 30044415

Durrer Schutz, D., Busetto, L., Dicker, D., Farpour-Lambert, N., Pryke, R., Toplak, H., Widmer, D., Yumuk, V., & Schutz, Y. (2019). European practical and patient-centred guidelines for adult obesity management in primary care. *Obesity Facts*, *12*(1), 40–66. DOI: 10.1159/000496183 PMID: 30673677

Effendi, A., Maulana, A., & Dani, D. (2023). Effective Public Health Strategies for Preventing Non Communicable Diseases Through Risk Factor Management and Community Based Initiatives. *Journal of Asian-african Focus in Health*, *1*(2), 68–76. DOI: 10.71435/595694

Eskandar, K. (2023). Artificial intelligence in healthcare: Explore the applications of AI in various medical domains, such as medical imaging, diagnosis, drug discovery, and patient care. *Series Med Sci*, *4*, 37–53.

Everett, E., Kane, B., Yoo, A., Dobs, A., & Mathioudakis, N. (2018). A Novel Approach for Fully Automated, Personalized Health Coaching for Adults with Prediabetes: Pilot Clinical Trial [Article]. *Journal of Medical Internet Research, 20*(2), e72. DOI: 10.2196/jmir.9723 PMID: 29487046

Fernandez-Luque, L., Núñez-Benjumea, F. J., Cervera-Torres, S., López-Guerra, J. L., Liao, Z., Pagán, J. A., & Zambrana, F. (2022). The Role of Patient-Generated Data in Personalized Oncology Care and Research: Opportunities and Challenges for Real-World Implementation. *Personal Health Informatics: Patient Participation in Precision Health,* 171-197.

Gao, Z., & Lee, J. E. (2019). Emerging technology in promoting physical activity and health: Challenges and opportunities. *Journal of Clinical Medicine, 8*(11), 1830. DOI: 10.3390/jcm8111830 PMID: 31683951

Gómez-García, C. A., Askar-Rodriguez, M., & Velasco-Medina, J. (2021). Platform for healthcare promotion and cardiovascular disease prevention. *IEEE Journal of Biomedical and Health Informatics, 25*(7), 2758–2767. DOI: 10.1109/JBHI.2021.3051967 PMID: 33449888

Gonzalez-Hernandez, G., Sarker, A., O'Connor, K., & Savova, G. (2017). Capturing the patient's perspective: A review of advances in natural language processing of health-related text. *Yearbook of Medical Informatics, 26*(01), 214–227. DOI: 10.15265/IY-2017-029 PMID: 29063568

Graue, M., Igland, J., Oftedal, B. F., Haugstvedt, A., Riise, H. K. R., Zoffmann, V., Jenum, A. K., Richards, D., & Kolltveit, B.-C. H. (2024). Interprofessional follow-up for people at risk of type 2 diabetes in primary healthcare - a randomized controlled trial with embedded qualitative interviews. *Scandinavian Journal of Primary Health Care, 42*(3), 450–462. https://doi.org/https://dx.doi.org/10.1080/02813432.2024.2337071. DOI: 10.1080/02813432.2024.2337071 PMID: 38588447

Haider, R., Sudini, L., Chow, C. K., & Cheung, N. W. (2019). Mobile phone text messaging in improving glycaemic control for patients with type 2 diabetes mellitus: A systematic review and meta-analysis. *Diabetes Research and Clinical Practice, 150*, 27–37. DOI: 10.1016/j.diabres.2019.02.022 PMID: 30822496

Han, C. Y., Lim, S. L., Ong, K. W., Johal, J., & Gulyani, A. (2024). Behavioral Lifestyle Intervention Program Using Mobile Application Improves Diet Quality in Adults With Prediabetes (D'LITE Study): A Randomized Controlled Trial. *Journal of the Academy of Nutrition and Dietetics, 124*(3), 358–371. DOI: 10.1016/j.jand.2023.10.005 PMID: 37820787

Hsiao, W. W.-W., Lin, J.-C., Fan, C.-T., & Chen, S. S.-S. (2022). Precision health in Taiwan: A data-driven diagnostic platform for the future of disease prevention. *Computational and Structural Biotechnology Journal, 20*, 1593–1602. DOI: 10.1016/j.csbj.2022.03.026 PMID: 35495110

Ibrahim, M. S., Mohamed Yusoff, H., Abu Bakar, Y. I., Thwe Aung, M. M., Abas, M. I., & Ramli, R. A. (2022). Digital health for quality healthcare: A systematic mapping of review studies. *Digital Health, 8*, 20552076221085810. DOI: 10.1177/20552076221085810 PMID: 35340904

Jakob, R., Harperink, S., Rudolf, A. M., Fleisch, E., Haug, S., Mair, J. L., Salamanca-Sanabria, A., & Kowatsch, T. (2022). Factors influencing adherence to mHealth apps for prevention or management of noncommunicable diseases: Systematic review. *Journal of Medical Internet Research, 24*(5), e35371. DOI: 10.2196/35371 PMID: 35612886

Khandakar, S., Al Mamun, M., Islam, M., Hossain, K., Melon, M., & Javed, M. (2024). Unveiling early detection and prevention of cancer: Machine learning and deep learning approaches. *Educational Administration: Theory and Practice, 30*(5), 14614–14628.

Kim, M. T., Kim, K. B., Nguyen, T. H., Ko, J., Zabora, J., Jacobs, E., & Levine, D. (2019). Motivating people to sustain healthy lifestyles using persuasive technology: A pilot study of Korean Americans with prediabetes and type 2 diabetes. *Patient Education and Counseling, 102*(4), 709–717. DOI: 10.1016/j.pec.2018.10.021 PMID: 30391298

Krist, A. H., Tong, S. T., Aycock, R. A., & Longo, D. R. (2017). Engaging patients in decision-making and behavior change to promote prevention. *Information Services & Use, 37*(2), 105–122. DOI: 10.3233/ISU-170826 PMID: 28972524

Kronlid, C., Brantnell, A., Elf, M., Borg, J., & Palm, K. (2024). Sociotechnical analysis of factors influencing IoT adoption in healthcare: A systematic review. *Technology in Society, 78*, 102675. DOI: 10.1016/j.techsoc.2024.102675

Larson, H. J., Lin, L., & Goble, R. (2022). Vaccines and the social amplification of risk. *Risk Analysis, 42*(7), 1409–1422. DOI: 10.1111/risa.13942 PMID: 35568963

Lawrence, K., Rodriguez, D. V., Feldthouse, D. M., Shelley, D., Yu, J. L., Belli, H. M., Gonzalez, J., Tasneem, S., Fontaine, J., Groom, L. L., Luu, S., Wu, Y., McTigue, K. M., Rockette-Wagner, B., & Mann, D. M. (2021). Effectiveness of an integrated engagement support system to facilitate patient use of digital diabetes prevention programs: Protocol for a randomized controlled trial [Article]. *JMIR Research Protocols*, *10*(2), e26750. Advance online publication. DOI: 10.2196/26750 PMID: 33560240

Lim, J., Hong, M., Lam, W., Zhang, Z., Teo, Z., Liu, Y., Ng, W., Foo, L., & Ting, D. (2022). Novel technical and privacy-preserving technology for artificial intelligence in ophthalmology. *Current Opinion in Ophthalmology*, *33*(3), 174–187. DOI: 10.1097/ICU.0000000000000846 PMID: 35266894

Liu, V. X., Morehouse, J. W., Baker, J. M., Greene, J. D., Kipnis, P., & Escobar, G. J. (2016). Data that drive: Closing the loop in the learning hospital system. *Journal of Hospital Medicine*, *11*(S1), S11–S17. DOI: 10.1002/jhm.2651 PMID: 27805797

Madimenos, F. C., Gildner, T. E., Eick, G. N., Sugiyama, L. S., & Snodgrass, J. J. (2022). Bringing the lab bench to the field: Point-of-care testing for enhancing health research and stakeholder engagement in rural/remote, indigenous, and resource-limited contexts. *American Journal of Human Biology*, *34*(11), e23808. DOI: 10.1002/ajhb.23808 PMID: 36166487

Manchanda, N., Aggarwal, A., Setya, S., & Talegaonkar, S. (2022). Digital intervention for the management of Alzheimer's disease. *Current Alzheimer Research*, *19*(14), 909–932. DOI: 10.2174/1567205020666230206124155 PMID: 36744687

Martins, C., Godycki-Cwirko, M., Heleno, B., & Brodersen, J. (2018). Quaternary prevention: reviewing the concept: Quaternary prevention aims to protect patients from medical harm. *The European Journal of General Practice*, *24*(1), 106–111. DOI: 10.1080/13814788.2017.1422177 PMID: 29384397

Miller, A. S., Cafazzo, J. A., & Seto, E. (2016). A game plan: Gamification design principles in mHealth applications for chronic disease management. *Health Informatics Journal*, *22*(2), 184–193. DOI: 10.1177/1460458214537511 PMID: 24986104

Mundra, A., Kalantri, A., Jakasania, A., Sathe, H., Raut, A., Maliye, C., Bahulekar, P., Dawale, A., Paradkar, R. J., Siriah, S., Kumar, S., Gupta, S. S., & Garg, B. (2023). Vitalizing community for health promotion against modifiable risk factors of noncommunicable diseases (V-CaN) in rural central India: Protocol for a hybrid type II implementation effectiveness trial. *JMIR Research Protocols*, *12*(1), e42450. DOI: 10.2196/42450 PMID: 37773622

Nahum-Shani, I., & Almirall, D. (2019). An Introduction to Adaptive Interventions and SMART Designs in Education. *NCSER 2020-001*. National center for special education research.

Neto, O. L., & Von Wyl, V. (2024). Digital transformation of public health for noncommunicable diseases: Narrative viewpoint of challenges and opportunities. *JMIR Public Health and Surveillance*, *10*(1), e49575. DOI: 10.2196/49575 PMID: 38271097

Nguyen, V., Ara, P., Simmons, D., & Osuagwu, U. L. (2024). The Role of Digital Health Technology Interventions in the Prevention of Type 2 Diabetes Mellitus: A Systematic Review. *Clinical Medicine Insights. Endocrinology and Diabetes*, *17*, 1–11. DOI: 10.1177/11795514241246419 PMID: 38779330

Novak, J., Jurkova, K., Lojkaskova, A., Jaklova, A., Kuhnova, J., Pfeiferova, M., Kral, N., Janek, M., Omcirk, D., Malisova, K., Maes, I., Dyck, D. V., Wahlich, C., Ussher, M., Elavsky, S., Cimler, R., Pelclova, J., Tufano, J. J., Steffl, M., & Vetrovsky, T. (2024). Participatory development of an mHealth intervention delivered in general practice to increase physical activity and reduce sedentary behaviour of patients with prediabetes and type 2 diabetes (ENERGISED). *BMC Public Health*, *24*(1), 1–15. DOI: 10.1186/s12889-024-18384-2 PMID: 38556892

Nurul Fatihah Mohd, F., Sharifah Wajihah Wafa Syed Saadun Tarek, W., Naresh Bhaskar, R., Mohd Ibrahim, A., Norkhairani Abdul, R., Nurulhuda Mat, H., & Rohayah, H. (2023). Diabetes prevention through digital therapy for high-risk individuals: Study protocol for the Malaysia Diabetes Prevention Programme (MyDiPP). *Malaysian Journal of Nutrition*, *29*(3), 147–162. DOI: 10.31246/mjn-2022-0015

Orte, S., Migliorelli, C., Sistach-Bosch, L., Gómez-Martínez, M., & Boqué, N. (2023). A tailored and engaging mHealth gamified framework for nutritional behaviour change. *Nutrients*, *15*(8), 1950. DOI: 10.3390/nu15081950 PMID: 37111168

Park, S., Garcia-Palacios, J., Cohen, A., & Varga, Z. (2019). *From treatment to prevention: The evolution of digital healthcare*. Bayer Pharmaceuticals, 178 Muellerstrasse, 13353 Berlin, Germany.

Razzak, M. I., Imran, M., & Xu, G. (2020). Big data analytics for preventive medicine. *Neural Computing & Applications*, *32*(9), 4417–4451. DOI: 10.1007/s00521-019-04095-y PMID: 32205918

Sarfati, D., McLeod, M., Stanley, J., Signal, V., Stairmand, J., Krebs, J., Dowell, A., Leung, W., Davies, C., & Grainger, R. (2018). BetaMe: impact of a comprehensive digital health programme on HbA1c and weight at 12 months for people with diabetes and pre-diabetes: study protocol for a randomised controlled trial. *Trials, 19*(1), 161. https://doi.org/https://dx.doi.org/10.1186/s13063-018-2528-4. DOI: 10.1186/s13063-018-2528-4 PMID: 29506562

Sartelli, M. C., Hardcastle, T., Catena, F., Chichom-Mefire, A., Coccolini, F., Dhingra, S., Haque, M., Hodonou, A., Iskandar, K., & Labricciosa, F. M. (2020). Antibiotic use in low and middle-income countries and the challenges of antimicrobial resistance in surgery. *Antibiotics (Basel, Switzerland), 9*(8), 497. DOI: 10.3390/antibiotics9080497 PMID: 32784880

Schueller, S. M., Tomasino, K. N., Lattie, E. G., & Mohr, D. C. (2016). Human support for behavioral intervention technologies for mental health: The efficiency model. *Assoc Comput Machin, 21*, 22.

Shahmoradi, L., Safadari, R., & Jimma, W. (2017). Knowledge management implementation and the tools utilized in healthcare for evidence-based decision making: A systematic review. *Ethiopian Journal of Health Sciences, 27*(5), 541–558. DOI: 10.4314/ejhs.v27i5.13 PMID: 29217960

Shahzad, S. K., Ahmed, D., Naqvi, M. R., Mushtaq, M. T., Iqbal, M. W., & Munir, F. (2021). Ontology driven smart health service integration. *Computer Methods and Programs in Biomedicine, 207*, 106146. DOI: 10.1016/j.cmpb.2021.106146 PMID: 34020375

Sharma, N., & Kaushik, P. (2025). Integration of AI in Healthcare Systems—A Discussion of the Challenges and Opportunities of Integrating AI in Healthcare Systems for Disease Detection and Diagnosis. *AI in Disease Detection: Advancements and Applications*, 239-263.

Summers, C., Tobin, S., & Unwin, D. (2021). Evaluation of the low carb program digital intervention for the self-management of type 2 diabetes and prediabetes in an nhs england general practice: Single-arm prospective study [Article]. *JMIR Diabetes, 6*(3), e25751. Advance online publication. DOI: 10.2196/25751 PMID: 34499035

Swinckels, L., Bennis, F. C., Ziesemer, K. A., Scheerman, J. F., Bijwaard, H., de Keijzer, A., & Bruers, J. J. (2024). The use of deep learning and machine learning on longitudinal electronic health records for the early detection and prevention of diseases: Scoping review. *Journal of Medical Internet Research, 26*, e48320. DOI: 10.2196/48320 PMID: 39163096

Taylor, R. S., Dalal, H. M., & McDonagh, S. T. (2022). The role of cardiac rehabilitation in improving cardiovascular outcomes. *Nature Reviews. Cardiology*, *19*(3), 180–194. DOI: 10.1038/s41569-021-00611-7 PMID: 34531576

Teoh, K. W., Ng, C. M., Chong, C. W., Cheong, W. L., Ng, Y. L., Bell, J. S., & Lee, S. W. H. (2024). A digital health-supported and community pharmacy-based lifestyle intervention program for adults with pre-diabetes: A study protocol for a cluster randomised controlled trial [Article]. *BMJ Open*, *14*(10), e083921. DOI: 10.1136/bmjopen-2024-083921 PMID: 39448216

Tesser, C. D., & Norman, A. H. (2021). Quaternary prevention and medicalisation: Inseparable concepts. *Interface: Comunicacao, Saude, Educacao*, *25*, e210101. DOI: 10.1590/interface.210101

Toro-Ramos, T., Michaelides, A., Anton, M., Karim, Z., Kang-Oh, L., Argyrou, C., Loukaidou, E., Charitou, M. M., Sze, W., & Miller, J. D. (2020). Mobile delivery of the diabetes prevention program in people with prediabetes: Randomized controlled trial [Article]. *JMIR mHealth and uHealth*, *8*(7), e17842. Advance online publication. DOI: 10.2196/17842 PMID: 32459631

Vetrovsky, T., Kral, N., Pfeiferova, M., Kuhnova, J., Novak, J., Wahlich, C., Jaklova, A., Jurkova, K., Janek, M., Omcirk, D., Capek, V., Maes, I., Steffl, M., Ussher, M., Tufano, J. J., Elavsky, S., Van Dyck, D., Cimler, R., Yates, T., & Seifert, B. (2023). mHealth intervention delivered in general practice to increase physical activity and reduce sedentary behaviour of patients with prediabetes and type 2 diabetes (ENERGISED): Rationale and study protocol for a pragmatic randomised controlled trial. *BMC Public Health*, *23*(1), 1–16. DOI: 10.1186/s12889-023-15513-1 PMID: 36997936

Waldman, S. A., & Terzic, A. (2019). Healthcare evolves from reactive to proactive. *Clinical Pharmacology and Therapeutics*, *105*(1), 10–13. DOI: 10.1002/cpt.1295 PMID: 30597564

Weedon-Fekjær, H., Romundstad, P. R., & Vatten, L. J. (2014). Modern mammography screening and breast cancer mortality: Population study. *BMJ (Clinical Research Ed.)*, *348*(jun17 17), 348. DOI: 10.1136/bmj.g3701 PMID: 24951459

Weitzel, J. N., Blazer, K. R., MacDonald, D. J., Culver, J. O., & Offit, K. (2011). Genetics, genomics, and cancer risk assessment: State of the art and future directions in the era of personalized medicine. *CA: a Cancer Journal for Clinicians*, *61*(5), 327–359. DOI: 10.3322/caac.20128 PMID: 21858794

Chapter 11
Acute Care

ABSTRACT

This chapter explores the transformative role of Knowledge Management (KM) and digital health solutions in enhancing acute care delivery within high-stakes settings like emergency departments and ICUs. Amid evolving patient needs and technological advancements, acute care demands rapid, precise decision-making under time constraints. Digital tools AI-powered Clinical Decision Support Systems (CDSSs), predictive analytics, and blockchain-enabled knowledge repositories integrate multimodal data (EHRs, sensors, genomics) to optimize clinical outcomes, reduce errors, and personalize care. However, challenges such as interoperability, clinician resistance, and ethical concerns (e.g., AI bias) persist.

INTRODUCTION

The healthcare sector has undergone a remarkable evolution over the past decade, driven by technological innovation and the increasing complexity of patient needs. Within this evolving landscape, acute care settings such as emergency departments, intensive care units (ICUs), surgical units, trauma care, and acute medical units emerge as some of the most challenging and knowledge-intensive domains of modern healthcare delivery. Festila and Müller (2022) demonstrate that critical care environments, including specialties like anaesthesiology and pain medicine, integrate heterogeneous technologies that frequently disrupt workflows, necessitating continuous adaptation, coordination, and realignment of practices among healthcare professionals to maintain effective patient care. Acute care is engineered to address severe illnesses or injuries requiring immediate intervention, characterized by decision making under strict time constraints and high-risk environment. This demands a strong relationship of technical expertise (e.g., interpreting real-time physiological data) and non-technical skills (e.g., team communication), alongside seamless collaboration across multidisciplinary teams and resource domains (Cascella

DOI: 10.4018/979-8-3373-1877-6.ch011

et al., 2023). Thus, there is a growing emphasis on personalized care, and the rapid expansion of health data from electronic health records (EHRs), wearable sensors, and genomic datasets highlight the critical need for advanced Knowledge Management (KM) frameworks integrated with digital health solutions. These frameworks are vital for optimizing clinical decision-making, minimizing medical errors, and enhancing patient outcomes in these dynamic settings (Alhashmi, 2025).

This chapter investigates how KM, underpinned by digital technologies, can transform acute care delivery more effectively. Digital health solutions such as AI-powered CDSSs that analyze EHR and sensor data for real-time recommendations, predictive models forecasting patient deterioration, and blockchain-enabled repositories for secure knowledge sharing offer a robust infrastructure for intelligent healthcare operations. For instance, AI algorithms can process multimodal data (e.g., vital signs, lab results) to suggest precise interventions, while blockchain ensures traceability of shared clinical insights without compromising patient privacy. Nevertheless, their integration faces barriers, including interoperability issues between legacy systems and modern platforms, clinician resistance due to perceived autonomy loss, alert fatigue from frequent notifications, and ethical concerns over AI accountability and bias (Reeves et al., 2017). Addressing these requires a KM strategy that aligns technological capabilities with human factors and organizational governance.

The subsequent subtopics of this chapter are organized as follows. Section 11.2 examines the role of digital health solutions in supporting acute care, highlighting their capabilities and applications. Section 11.3 explores the importance of KM tools, techniques, and technologies in enabling personalized acute care delivery. Section 11.4 provides the case examples and section 11.5 presents clinical decision making (CLOTS) framework, to illustrate the practical impact of these approaches on clinical decision-making. Section 11.6 concludes with a chapter summary, synthesizing key insights and their implications for intelligent healthcare operations.

ROLE OF DIGITAL HEALTH SOLUTIONS TO SUPPORT ACUTE CARE

Digital health solutions have emerged as transformative tools in acute care settings, addressing the escalating demands of clinical efficiency, decision-making precision, and care coordination in high-stakes environments. As outlined above, acute care spanning emergency departments, intensive care units (ICUs), surgical units, trauma care, and acute medical units requires rapid responses to severe conditions under strict time constraints. This section explores how digital technologies support these demands, building on the evolution from traditional workflows to integrated, intelligent systems.

Digital Transformation in Acute Care Settings

A robust body of literature highlights the shift from paper-based to digital workflows in acute and intensive care, emphasizing improvements in efficiency, coordination, and clinical outcomes (Watson, 2016; Yilmaztürk et al., 2023). For instance, Watson (2016) reports that digitizing ICU nursing care forms reduced documentation time by approximately 57 minutes per patient per day, yielding substantial time and cost savings. This transition reflects a broader digital transformation, driven by tools such as EHRs, telemedicine, and real-time analytics, which have become integral to modern acute care delivery. Understanding their evolution and impact is critical for navigating the future of healthcare transformation, particularly in settings where delays or errors can have life-altering consequences.

The progression of digital health solutions in acute care can be categorized into three distinct generations:

- First-Generation Systems (1990s–2000s): These systems introduced basic EHRs as digital repositories, replacing paper-based records with static documentation. Their primary function was data storage, offering limited interactivity or decision support.
- Second-Generation Systems (2000s–2010s): Building on earlier foundations, these systems incorporated advanced functionalities such as computerized provider order entry (CPOE), results reporting, and rudimentary clinical decision support. They enabled more dynamic interactions but remained siloed within specific workflows.
- Third-Generation Systems (2010s–Present): Current systems represent integrated digital ecosystems, leveraging advanced analytics, AI, interoperability across care settings, and patient engagement tools. These platforms transcend static documentation, actively supporting clinical reasoning and interprofessional collaboration.

This evolution to third-generation systems marks a paradigm shift in acute care delivery. Unlike their predecessors, these technologies function as dynamic platforms that integrate multimodal data—hospital records, Internet of Things (IoT) device outputs, and genomic insights into actionable intelligence. In complex environments like ICUs or end-of-life care, they facilitate collaborative decision-making and collective sense-making, empowering healthcare professionals, including nurses and allied staff, to serve as information brokers and patient advocates (Adams et al., 2011; Donnelly et al., 2021). For example, real-time data streaming from ventilators or cardiac monitors, processed through AI algorithms, enables rapid adjustments to treatment plans, enhancing responsiveness in critical scenarios.

Modern digital health solutions extend beyond data management to enable real-time decision-making, workflow optimization, and predictive risk assessment. The big data in healthcare sourced from EHRs, patient histories, diagnostics, and IoT devices has fuelled the development of AI-powered Clinical Decision Support Systems (CDSSs). These systems harness genomic and biomedical research insights to deliver tailored treatment recommendations, easing cognitive overload for clinicians. Predictive analytics, supported by high-performance computing, can forecast critical events such as sepsis or stroke, enabling pre-emptive interventions that improve patient survival rates (Dash et al., 2019). For instance, predictive models analyzing vital sign trends have reduced sepsis mortality by up to 20% in some ICU settings by triggering alerts 6–12 hours before clinical deterioration becomes evident.

To illustrate the scope of these technologies, Table 1 provides an overview of key digital health solutions deployed in acute care, adapted and expanded from the original framework:

Table 1. Overview of key digital health solutions

Technology	Primary Functions	Benefits	Implementation Challenges	Interprofessional Considerations
Advanced EHRs	Comprehensive clinical documentation, order management, results reporting, care coordination	Improved information access, reduced documentation time, enhanced communication	Implementation complexity, user resistance, interoperability issues	Requires standardized documentation practices across professions
CDSSs	Real-time guidance based on patient data, clinical guidelines, and evidence-based protocols	Reduced errors, improved adherence to best practices, standardized care delivery	Alert fatigue, integration with workflow, keeping knowledge base current	Must support diverse clinical reasoning models-risk of privileging certain professional perspectives
Predictive Analytics	Identification of deteriorating patients, resource forecasting, capacity management	Earlier intervention, optimized resource allocation, improved throughput	Data quality issues, algorithm transparency, clinician trust	Opportunity to highlight diverse contributors to patient outcomes-supports proactive team mobilization
Telehealth & Virtual Care	Remote consultation, virtual rounds, home monitoring, family engagement	Extended care capacity, improved access, enhanced care transitions	Technical reliability, digital equity, licensure issues	Enables new models of interprofessional collaboration- requires adaptation of team communication practices

continued on following page

Table 1. Continued

Technology	Primary Functions	Benefits	Implementation Challenges	Interprofessional Considerations
Mobile Clinical Communication	Secure messaging, clinical photography, voice communication, team coordination	Improved team coordination, reduced delays, enhanced handoffs	Device management, information security, interruption management	Enables effective team communication-requires new norms for urgent vs. non-urgent communication
Computer Vision & AI	Automated image interpretation, activity monitoring, anomaly detection	Accelerated diagnosis, reduced cognitive load, enhanced surveillance	Algorithm validation, regulatory approval, clinician acceptance	May change professional roles and responsibilities-creates need for shared understanding of AI capabilities and limitations

This framework emphasizes interprofessional considerations, recognizing that digital solutions must support team-based care rather than isolated workflows. However, implementation experiences vary widely across settings, complicating generalizations about effectiveness. Organizations adopting a strategic, integrated approach rather than piecemeal adoption tend to realize greater benefits, as isolated implementations often fail to address systemic workflow gaps (Butt, 2020).

Three distinct approaches to digital transformation in acute care have emerged:

1. Technology-Driven Approach: Prioritizes system deployment with minimal focus on workflow redesign or KM, often leading to resistance and suboptimal outcomes.
2. Workflow-Driven Approach: Centers on process improvement, using technology as an enabler, enhancing efficiency but potentially overlooking knowledge-sharing needs.
3. Knowledge-Driven Approach: Focuses on capturing, sharing, and applying clinical knowledge, with technology and workflows aligned to this goal, fostering adaptability and collaboration.

Figure 1 illustrates the comparative impact of digital transformation approaches in acute care would depict how the knowledge-driven approach yields superior outcomes in decision-making accuracy and team coordination compared to technology- or workflow-driven models, based on metrics like error rates and response times.

Figure 1. Comparative impact of digital transformation approaches in acute care

APPROACHES TO DIGITAL TRANSFORMATION		
TECHNOLOGY-DRIVEN	**WORKFLOW-DRIVEN**	**KNOWLEDGE-DRIVEN**
Focus: System implementation	Focus: Process improvement	Focus: Clinical knowledge
Technology as the primary goal	Technology as enabler of better processes	Technology as platform for knowledge work
COMPARATIVE OUTCOMES		
Moderate initial efficiency gains	Improved workflow efficiency	Enhanced clinical decision quality
Limited clinical impact	Moderate clinical impact	Significant clinical outcome improvement
High user resistance	Moderate user satisfaction	High user satisfaction
Poor ROI over time	Moderate long-term value	Strong ROI and sustained improvements
Limited interprofessional impact	Some interprofessional collaboration improvements	Significant enhancement of team-based care delivery

Integration Challenges and Approaches

While individual digital health technologies have been extensively studied, comprehensive analyses of their integration into complex acute care environments are still limited. This subsection addresses these gaps, drawing on case studies and proposing KM-driven solutions to enhance adoption.

Case Study: Integration Challenges in NHS Acute Hospitals

While individual digital health technologies have been widely studied, comprehensive research addressing the integration challenges faced by acute care organizations when implementing these technologies in complex clinical environments remains limited. Based on case studies from Northumbria Healthcare, Sheffield Teaching Hospitals, Airedale NHS Foundation Trust, Yeovil District Hospital, and South Warwickshire NHS Foundation Trust, five key integration challenges have been identified (Naylor et al., 2015).

One of the most pressing challenges is technical integration, where hospitals struggle with inconsistent electronic health record (EHR) systems and data-sharing limitations between acute hospitals and community care. For example, Northumbria Healthcare NHS Foundation Trust faced difficulties integrating community-based care data with hospital records, resulting in gaps in continuity of care. Airedale NHS Foundation Trust addressed this by adopting telehealth systems, allowing care home staff to access hospital-based support, leading to a 37% reduction in avoidable admissions and a 45% decrease in A&E attendances.

Workflow integration presents another significant barrier, as many hospitals face difficulties embedding digital solutions into clinical workflows without causing disruption. South Warwickshire NHS Foundation Trust introduced the Discharge to Assess model to improve patient transitions from hospital to community care. This approach streamlined discharge planning and ensured patients received timely support, reducing unnecessary delays. Despite this, hospitals still face challenges in fully aligning discharge processes with digital tracking systems and integrating community services efficiently.

Cognitive integration refers to the challenge of presenting clinical information in a way that supports, rather than complicates, decision-making. Clinicians often struggle with fragmented information, making real-time data access difficult. Sheffield Teaching Hospitals' Right First-Time initiative improved data accessibility, leading to a decline in hospital bed usage among high-risk patients. The implementation of clinical dashboards played a crucial role in this effort by providing real-time access to patient records across multiple care settings.

At an organizational level, integration challenges emerge from difficulties in aligning digital and operational strategies with broader health system goals. Many hospitals face obstacles in coordinating with primary care due to separate governance structures and funding models. Northumbria Healthcare NHS Foundation Trust addressed this issue by establishing multi-agency governance structures that ensured shared accountability between hospitals and community services. Such joint leadership models enabled acute hospitals to play a more active role in community-based care, fostering a more cohesive healthcare system.

Finally, team integration remains a key concern, as poor communication between primary care providers, social workers, and hospital teams often leads to fragmented care. Airedale NHS Foundation Trust introduced multidisciplinary telehealth consultations, which helped reduce hospital reliance for elderly patients by enabling virtual collaboration between healthcare professionals across different settings. The implementation of virtual multidisciplinary teams (MDTs) further improved interprofessional collaboration and reduced hospital dependency, allowing patients to receive better-coordinated care.

Table 2. KM-driven integration framework for digital health in acute care

Integration Dimension	KM Focus	Challenges	KM-Driven Solutions
Technical Integration	Ensuring seamless knowledge exchange across systems through interoperability and data standardization.	EHR interoperability gaps limiting continuity of knowledge and data sharing.	Adoption of AI-enhanced knowledge repositories, FHIR-based APIs for interoperability.
Workflow Integration	Embedding KM frameworks into clinical workflows to enable real-time knowledge access and process optimization.	Lack of real-time knowledge access affecting discharge planning and workflow efficiency.	Integration of AI-driven process automation, real-time clinical dashboards for workflow management.
Cognitive Integration	Presenting knowledge in clinically meaningful ways to support decision-making and reduce cognitive overload.	Cognitive overload due to fragmented clinical data and poor data visualization.	Development of knowledge visualization tools, AI-driven clinical decision support (CDSS).
Organizational Integration	Aligning knowledge governance with hospital strategies to ensure a cohesive and system-wide KM approach.	Misalignment between acute and primary care knowledge systems, leading to inefficient knowledge governance.	Implementation of joint KM leadership models, centralized KM platforms for healthcare networks.
Team Integration	Enhancing knowledge exchange across disciplines to foster collaboration and interprofessional learning.	Siloed communication and restricted knowledge sharing among interdisciplinary teams.	Use of virtual multidisciplinary teams (MDTs), collaborative knowledge-sharing platforms.

This framework leverages KM to address integration challenges, aligning with the knowledge-driven approach. For instance, AI-enhanced repositories can standardize data exchange, while virtual MDTs foster team-based learning, reducing silos and enhancing personalized care delivery.

The integration of digital health solutions in acute care offers transformative potential but requires strategic alignment with KM principles. Technical solutions alone are insufficient; workflow redesign and knowledge-sharing mechanisms are critical for success. Future implementations should prioritize interoperability standards (e.g., FHIR) and clinician training to mitigate resistance and ethical concerns, ensuring these technologies enhance, rather than disrupt, acute care operations.

IMPORTANCE OF KM TOOLS, TECHNIQUES, AND TECHNOLOGIES FOR SUPPORTING PERSONALIZED ACUTE CARE DELIVERY

Acute care settings, characterized by high-pressure conditions and time-sensitive decision-making, demand real-time, evidence-based responses to complex patient needs (Zaboli et al., 2025). As highlighted in above, these environments surrounding emergency departments, ICUs, and surgical units require seamless coordination and rapid adaptation amidst evolving clinical knowledge and multidisciplinary collaboration. Effective KM systems are indispensable for optimizing patient outcomes, reducing medical errors, and enhancing operational efficiency in such contexts. This section explores how KM tools, techniques, and technologies enable personalized acute care delivery.

KM Frameworks for Personalized Acute Care

KM systems provide a structured approach to capturing, organizing, and applying clinical knowledge, ensuring healthcare professionals can access critical insights at the point of care. In an era where personalized medicine is a cornerstone of healthcare, KM tools augmented by Artificial Intelligence (AI) and Machine Learning (ML) enable tailored interventions based on individual patient data—clinical histories, real-time physiological metrics, genetic markers, and risk profiles. For instance, intelligent case-based KM systems (ICKMS) integrate techniques like fuzzy logic and case-based reasoning to assess care quality, identify performance gaps, and deliver customized retraining, significantly improving service standards and reducing complaints, as demonstrated in long-term care settings (Choy et al., 2018). In acute care, these systems adapt to dynamic conditions, supporting precision medicine by aligning treatments with patient-specific disease trajectories (Kandhare et al., 2025).

AI-driven KM tools, such as CDSSs, enhance decision-making by analyzing patient data against clinical guidelines and predictive models. These systems reduce variability, improve diagnostic accuracy, and promote adherence to evidence-based practices, mitigating the cognitive burden on clinicians (Mitra & Rehman, 2025). For example, in kidney transplant care, AI-driven chatbots provide real-time access to guidelines, personalized education, and medication management recommendations, shifting from generalized protocols to patient-specific interventions (Garcia Valencia et al., 2023). Similarly, advanced EHRs with KM capabilities serve as centralized repositories, integrating structured data, clinical pathways, and historical insights. This longitudinal tracking enables clinicians to make informed decisions, such as adjusting therapies based on past responses, enhancing continuity and precision (Reeves et al., 2017).

Predictive analytics platforms further amplify KM's impact by using data-driven risk stratification to identify deteriorating patients and optimize treatment pathways. By analyzing trends in vital signs or lab results, these tools support early interventions, reducing adverse events like sepsis or cardiac arrest (Stone et al., 2022). In ICUs, predictive models can flag at-risk patients' hours before clinical symptoms manifest, enabling proactive care planning that aligns with personalized medicine goals. These capabilities highlight KM's role in transforming raw data into actionable knowledge.

Advanced KM Techniques and Technologies

The integration of KM into acute care relies on a suite of advanced techniques and technologies designed to enhance knowledge capture, retrieval, and application. Natural Language Processing (NLP) and AI-based thematic modelling extract structured insights from unstructured sources clinical notes, guidelines, and literature delivering context-aware recommendations (Kousis & Tjortjis, 2023). For instance, NLP can synthesize a patient's narrative history with diagnostic data to highlight nuanced factors (e.g., social determinants), reducing cognitive overload and supporting complex decision-making. This capability is critical in acute settings, where clinicians must assess multiple knowledge streams under time pressure.

Distributed Decision Support Systems (DDSSs) with AI-enhanced knowledge retrieval provide another layer of support. Leveraging data from Internet of Medical Things (IoMT) devices and case-based reasoning, DDSSs deliver context-specific insights rapidly, minimizing delays (Almadani et al., 2025). For example, in trauma care, a DDSS might integrate real-time vitals from a wearable sensor with historical case data to recommend an optimal intervention, enhancing timeliness and precision. Real-time CDSS integration, guided by frameworks like Simon's decision-making model, further refines this process. By aligning recommendations with individual risk profiles and disease trajectories via AI and advanced analytics, these systems enable precision medicine in dynamic scenarios, such as adjusting dosages during a patient's acute decline (Mosavi & Santos, 2024).

Collaborative knowledge-sharing mechanisms are equally vital. Interdisciplinary case discussions, AI-assisted diagnostic forums, and continuous learning modules powered by large language models (LLMs) disseminate critical insights across teams, improving accuracy and safety (Zhou et al., 2023). In an ICU, an AI-assisted forum might analyze a patient's data and prompt a multidisciplinary team to refine a ventilator strategy, reducing diagnostic errors. Interprofessional learning platforms, meanwhile, foster dynamic team learning by bridging disciplinary silos, enabling clinicians, nurses, and allied professionals to adapt to evolving contexts (Yeo, 2020).

These tools ensure that the latest evidence-based knowledge is accessible, aligning with team-based care.

Optimizing Personalized Care Pathways

ML significantly enhances KM-driven personalization by synthesizing data from EHRs and clinical sources to detect patient-specific patterns. Algorithms such as decision trees, support vector machines, and neural networks dynamically adapt treatment plans in real time, aligning with precision medicine objectives (Ahmed et al., 2020). For instance, in stroke care, ML models might analyze imaging and vitals to recommend a thrombolytic dose tailored to a patient's clot characteristics, improving efficacy and reducing costs. Data-driven decision intelligence integrates EHRs, administrative data, and research into a unified knowledge base, enhancing interdisciplinary communication and continuity (Rolla, 2023). When paired with CDSSs and computerized order entry, this approach improves safety, supports quality metrics, and enables community-level interventions, such as targeting high-risk populations for preventive care.

Blockchain-based KM systems offer a secure, decentralized solution for knowledge sharing across institutions. By leveraging immutable, access-controlled transactions, these systems ensure the trustworthiness of clinical data, ML models, and insights, critical for large hospital networks (Wijesekara & Gunawardena, 2023). For example, a regional collaboration might share validated sepsis prediction models via blockchain, maintaining privacy while enhancing coordination. This aligns with blockchain repositories, reinforcing KM's role in trusted, scalable care delivery.

Future Directions and Implications

Looking forward, KM will continue to evolve personalized acute care. AI-driven systems are poised to automate evidence synthesis, dynamically updating guidelines based on real-world outcomes. Self-learning CDSS networks will refine recommendations using clinician feedback and patient responses, enhancing adaptability (e.g., adjusting antibiotic regimens based on resistance patterns). Digital twin models, simulating patient scenarios, will enable clinicians to test treatment pathways virtually before implementation, offering a proactive approach to complex cases like multi-organ failure. These advancements promise to reduce errors, optimize resources, and align with the value-based healthcare vision.

However, challenges such as ethical concerns for example AI bias or accountability require robust governance. Future research should explore scalable KM frameworks that balance innovation with practicality, ensuring equitable access across diverse acute care settings.

CASE EXAMPLES

Case Example 1: A Race Against Time with Canon's AUTOStroke Solution

In early 2019, a busy U.S. stroke center grappled with a daily race: patients with acute ischemic stroke needed swift answers to unlock lifesaving thrombectomy. Radiologists, buried in Computed Tomography (CT) scans, often lost precious minutes spotting large vessel occlusions time that brain cells couldn't afford. That March, Canon's AUTOStroke Solution rolled out, an AI tool poised to shift the tide. Imagine a patient wheeled in, Computed Tomography Angiography (CTA) scan whirring; in under a minute, an alert flashed Large Vessel Occlusion (LVO) in the internal carotid artery. The stroke team leapt into gear, skipping the usual wait. Over the next year, this scene unfolded for hundreds, from Internal Carotid Artery (ICA) blockages to trickier M1 and M2 segments of the Middle Cerebral Artery (MCA) cases. The tool rarely misjudged a non-LVO, outpacing manual reads by far. For one patient, it meant surgery within hours, reviving blood flow and a future. Humming beside the CT, this AI turned scans into instant action, align with real-time analytics and decision support. Still, smaller M2 occlusions occasionally dodged detection, and rare contrast timing errors confused an M1 for an ICA. Yet, it armed the team to personalize care under pressure, reclaiming lives against the clock (Rava et al., 2021).

Case Example 2: Unmasking Sepsis with AI in Shanghai's ICUs

In the summer of 2021, the ICUs of Shanghai East and Renji Hospitals buzzed with urgency as patients arrived with telltale signs racing pulses, shallow breaths, fevers spiking or plummeting. Sepsis, a stealthy killer, loomed, but pinning it down fast enough to act was a struggle. Blood cultures lagged days behind, too slow for lives hanging by a thread. Enter a bold experiment, chronicled by Wang et al. (2024) in Clinica Chimica Acta: a ML tool paired with a new DNA sequencing trick to spot bacterial traces in blood plasma. One night, a patient's pressure crashed, his fever soared; within hours, the system lit up sepsis confirmed guiding antibiotics that turned the tide. Over two years, this played out across hundreds of beds, catching infections traditional tests missed. It sifted clues like procalcitonin and breathing rates, analytics, and fed real-time alerts to doctors via KM systems. For some, it was the edge between survival and collapse. Not every case was perfect tiny bacterial signals occasionally slipped through, and the tool craved broader proving grounds. Yet, in those high-stakes wards, it became a silent partner, personalizing care by unmasking sepsis fast. This wasn't just tech; it was a lifeline, blending AI smarts with human grit to rewrite outcomes in acute care's toughest battles (Wang et al., 2024).

CLINICAL DECISION MAKING (CLOTS)

Clinical decision-making lies at the heart of acute care delivery. The process of arriving at timely, evidence-based, and patient-centered decisions in high-stakes environments—such as emergency departments or ICUs is not only essential for improving patient outcomes, reducing medical errors, and optimizing resource use, but also deeply influenced by the physical and organizational design of the care environment, which must support clinical workflows and decision-making under pressure (Hamilton & Shepley, 2010). In the context of intelligent knowledge-based healthcare operations, the CLOTS framework clinical logic, operational tools, and systems represents a strategic approach to integrating digital health technologies, clinical reasoning, and KM tools for enhanced decision-making.

At its core, CLOTS emphasizes the synergy between three components: (1) Clinical Logic rooted in evidence-based practice, (2) Operational Tools such as AI-powered CDSS, and (3) interoperable Systems that consolidate data from EHRs, Internet of Medical Things (IoMT) devices, and other digital sources. This triad empowers clinicians to make rapid and informed decisions aligned with real-time patient data and contextual needs.

AI-enhanced CDSS now support clinicians by integrating vast amounts of structured and unstructured data including vital signs, diagnostic imaging, laboratory results, and genomic profiles to deliver personalized diagnostic and treatment recommendations, while also helping to mitigate cognitive biases and improve diagnostic accuracy in high-risk environments like emergency and critical care (Cerrato & Halamka, 2020). These systems reduce variability in clinical judgment and help flag early warning signs of deterioration, such as sepsis or stroke. For instance, ML algorithms have demonstrated the ability to detect sepsis several hours before clinical symptoms become apparent, enabling clinicians to initiate timely interventions that can significantly reduce morbidity, mortality, and associated healthcare costs (O'Reilly et al., 2024).

Real-time data integration through distributed decision support systems (DDSS) further strengthens the CLOTS model. As Bhushan et al. (2023) note, The global proliferation of the Internet of Medical Things (IoMT) has generated a deluge of health data, driven by widespread, interconnected sensing devices that support real-time monitoring and personalized care though their adoption is tempered by persistent challenges related to data security, privacy, and interoperability. DDSS solutions process this data at scale, leveraging predictive models and knowledge repositories to stratify risk, automate triage, and tailor interventions. Blockchain-based KM systems are emerging as secure and decentralized tools to manage knowledge sharing across healthcare networks particularly valuable in acute care collaborations between hospitals and regional care centers (Wijesekara & Gunawardena, 2023).

Additionally, Clinical decision-making is increasingly enhanced by NLP and AI-powered patient-centric knowledge graphs (PCKGs), which extract actionable insights from clinical notes, medical literature, and real-time communications offering a holistic, multidimensional view of patient health that supports more accurate diagnoses and personalized treatment planning (Al Khatib et al., 2024).

Importantly, the CLOTS framework acknowledges that digital tools are not replacements for clinical expertise but enhancers of collaborative decision-making. It supports interprofessional knowledge sharing through real-time communication platforms, AI-assisted forums, and embedded learning modules ensuring that decisions are informed by diverse perspectives. Despite its promise, the implementation of CLOTS-based systems must address key challenges. These include data silos, algorithmic bias, workflow integration, and the explainability of AI recommendations. Continuous training, robust KM infrastructures, and ethical governance frameworks are vital to sustaining clinical trust in intelligent systems. CLOTS represents a pragmatic and holistic model for operationalizing knowledge-based acute care. By combining advanced technologies with clinician intuition and interprofessional collaboration, it facilitates safer, faster, and more personalized decision-making in high-pressure settings.

CONCLUSION

The rapid evolution of healthcare, powered by technological innovation and complex patient needs, has placed acute care encompassing emergency departments, ICUs, and surgical units at the forefront of modern medicine. This chapter examines how KM and digital health solutions can transform these high-stakes environments, where swift, precise decisions are critical. It traces the shift from paper-based systems to advanced digital ecosystems, spotlighting tools like AI-driven CDSS, predictive analytics, and blockchain repositories that integrate data from EHRs, sensors, and genomics to enhance clinical outcomes and personalize care. Real-world examples, such as Canon's AUTOStroke for stroke detection and Shanghai's sepsis-spotting AI, showcase their impact, while the CLOTS framework illustrates a strategic blend of clinical logic, tools, and systems. Despite challenges like interoperability, clinician resistance, and ethical concerns, a KM-driven approach aligning technology with human factors promises to revolutionize acute care delivery.

REFERENCES

Adams, J. A., Bailey, D. E.Jr, Anderson, R. A., & Docherty, S. L. (2011). Nursing roles and strategies in end-of-life decision making in acute care: A systematic review of the literature. *Nursing Research and Practice, 2011*(1), 527834. DOI: 10.1155/2011/527834 PMID: 21994831

Al Khatib, H. S., Neupane, S., Kumar Manchukonda, H., Golilarz, N. A., Mittal, S., Amirlatifi, A., & Rahimi, S. (2024). Patient-centric knowledge graphs: A survey of current methods, challenges, and applications. *Frontiers in Artificial Intelligence, 7*, 1388479. DOI: 10.3389/frai.2024.1388479 PMID: 39540199

Alhashmi, S. M. (2025). Knowledge management, health data, and advanced data analytics to expedite solutions in the healthcare industry. In *Digital Healthcare, Digital Transformation and Citizen Empowerment in Asia-Pacific and Europe for a Healthier Society* (pp. 231–247). Elsevier. DOI: 10.1016/B978-0-443-30168-1.00005-0

Almadani, B., Kaisar, H., Thoker, I. R., & Aliyu, F. (2025). A Systematic Survey of Distributed Decision Support Systems in Healthcare. *Systems, 13*(3), 157. DOI: 10.3390/systems13030157

Bhushan, B., Kumar, A., Agarwal, A. K., Kumar, A., Bhattacharya, P., & Kumar, A. (2023). Towards a secure and sustainable internet of medical things (iomt): Requirements, design challenges, security techniques, and future trends. *Sustainability (Basel), 15*(7), 6177. DOI: 10.3390/su15076177

Butt, J. (2020). A conceptual framework to support digital transformation in manufacturing using an integrated business process management approach. *Designs, 4*(3), 17. DOI: 10.3390/designs4030017

Cascella, M., Cascella, A., Monaco, F., & Shariff, M. N. (2023). Envisioning gamification in anesthesia, pain management, and critical care: Basic principles, integration of artificial intelligence, and simulation strategies. *Journal of Anesthesia, Analgesia and Critical Care, 3*(1), 33. DOI: 10.1186/s44158-023-00118-2 PMID: 37697415

Cerrato, P., & Halamka, J. (2020). *Reinventing clinical decision support: data analytics, artificial intelligence, and diagnostic reasoning*. Taylor & Francis. DOI: 10.1201/9781003034339

Choy, K. L. T., Siu, K. Y. P., Ho, T. S. G., Wu, C., Lam, H. Y., Tang, V., & Tsang, Y. P. (2018). An intelligent case-based knowledge management system for quality improvement in nursing homes. *VINE Journal of Information and Knowledge Management Systems, 48*(1), 103–121. DOI: 10.1108/VJIKMS-01-2017-0001

Donnelly, S., Ó Coimín, D., O'Donnell, D., Ní Shé, É., Davies, C., Christophers, L., Mc Donald, S., & Kroll, T. (2021). Assisted decision-making and interprofessional collaboration in the care of older people: A qualitative study exploring perceptions of barriers and facilitators in the acute hospital setting. *Journal of Interprofessional Care, 35*(6), 852–862. DOI: 10.1080/13561820.2020.1863342 PMID: 33588668

Festila, M., & Müller, S. D. (2022). Coordinating knowledge work across technologies: Evidence from critical care practices. *Information and Organization, 32*(2), 100411. DOI: 10.1016/j.infoandorg.2022.100411

Garcia Valencia, O. A., Thongprayoon, C., Jadlowiec, C. C., Mao, S. A., Miao, J., & Cheungpasitporn, W. (2023). Enhancing kidney transplant care through the integration of chatbot. *Healthcare,* Hamilton, D. K., & Shepley, M. M. (2010). *Design for critical care: An evidence-based approach.* Routledge.

Kandhare, P., Kurlekar, M., Deshpande, T., & Pawar, A. (2025). A Review on Revolutionizing Healthcare Technologies with AI and ML Applications in Pharmaceutical Sciences. *Drugs and Drug Candidates, 4*(1), 9. DOI: 10.3390/ddc4010009

Kousis, A., & Tjortjis, C. (2023). Investigating the key aspects of a smart city through topic modeling and thematic analysis. *Future Internet, 16*(1), 3. DOI: 10.3390/fi16010003

Mosavi, N. S., & Santos, M. F. (2024). Enhancing Clinical Decision Support for Precision Medicine: A Data-Driven Approach. *Informatics,* Naylor, C., Alderwick, H., & Honeyman, M. (2015). *Acute hospitals and integrated care. From hospitals to health systems.* King's Fund (revista electrónica)[Internet], 1-90.

O'Reilly, D., McGrath, J., & Martin-Loeches, I. (2024). Optimizing artificial intelligence in sepsis management: Opportunities in the present and looking closely to the future. *Journal of Intensive Medicine, 4*(1), 34–45. DOI: 10.1016/j.jointm.2023.10.001 PMID: 38263963

Rava, R. A., Peterson, B. A., Seymour, S. E., Snyder, K. V., Mokin, M., Waqas, M., Hoi, Y., Davies, J. M., Levy, E. I., Siddiqui, A. H., & Ionita, C. N. (2021). Validation of an artificial intelligence-driven large vessel occlusion detection algorithm for acute ischemic stroke patients. *The Neuroradiology Journal, 34*(5), 408–417. DOI: 10.1177/1971400921998952 PMID: 33657922

Reeves, S., Palaganas, J., & Zierler, B. (2017). An updated synthesis of review evidence of interprofessional education. *Journal of Allied Health, 46*(1), 56–61. PMID: 28255597

Rolla, K. J. (2023). Trends and Futuristic Applications of Big Data and Electronic Health Record Data in Empowering Constructive Clinical Decision Support Systems. *Bio Science Research Bulletin-Biological Sciences,* 39(2).

Stone, K., Zwiggelaar, R., Jones, P., & Mac Parthaláin, N. (2022). A systematic review of the prediction of hospital length of stay: Towards a unified framework. *PLOS Digital Health, 1*(4), e0000017. DOI: 10.1371/journal.pdig.0000017 PMID: 36812502

Wang, L., Tian, W., Zhang, W., Wen, D., Yang, S., Wang, J., Han, X., Wang, J., Ding, W., Wang, L., Yu, Y., & Wu, W. (2024). A Machine learning model for predicting sepsis based on an optimized assay for microbial cell-free DNA sequencing. *Clinica Chimica Acta, 559*, 119716–119716. DOI: 10.1016/j.cca.2024.119716 PMID: 38710402

Watson, A. R. (2016). Impact of the digital age on transforming healthcare. *Healthcare information management systems: cases, strategies, and solutions,* 219-233.

Wijesekara, P. A. D. S. N., & Gunawardena, S. (2023). A review of blockchain technology in knowledge-defined networking, its application, benefits, and challenges. *Network (Bristol, England), 3*(3), 343–421.

Yeo, R. K. (2020). Crossing knowledge boundaries: From team learning to knowledge teams. *Small Group Research, 51*(6), 700–737. DOI: 10.1177/1046496420919929

Yilmaztürk, N., Kose, İ., & Cece, S. (2023). The effect of digitalization of nursing forms in ICUs on time and cost. *BMC Nursing, 22*(1), 201. DOI: 10.1186/s12912-023-01333-6 PMID: 37312143

Zaboli, A., Biasi, C., Magnarelli, G., Miori, B., Massar, M., Pfeifer, N., Brigo, F., & Turcato, G. (2025). Arterial Blood Gas Analysis and Clinical Decision-Making in Emergency and Intensive Care Unit Nurses: A Performance Evaluation. *Health Care.*

Zhou, H., Liu, F., Gu, B., Zou, X., Huang, J., Wu, J., Li, Y., Chen, S. S., Zhou, P., & Liu, J. (2023). A survey of large language models in medicine: Progress, application, and challenge. arXiv preprint arXiv:2311.05112.

Chapter 12
Rehab and Aged Care

ABSTRACT

The growing demands of an aging population and rehabilitation needs strain global healthcare systems, necessitating efficient, personalized care delivery. Knowledge Management (KM) addresses these challenges by systematically capturing, sharing, and applying clinical expertise. This chapter explores KM's role in enhancing rehabilitation and aged care through digital health innovations, such as Artificial Intelligence (AI)-driven Clinical Decision Support Systems (CDSS), tele-assessment platforms, and wearable technologies. These tools improve patient outcomes by tackling data overload and fragmented care, with computational simulations and case examples like dementia tele-assessment demonstrating their efficacy.

INTRODUCTION

The global healthcare landscape has undergone significant transformation in recent decades, driven by demographic shifts characterized by an increasingly aging population. This shift poses substantial challenges to healthcare systems worldwide, as documented by Jane Osareme et al. (2024). The rising number of older adults and patients requiring rehabilitation services, particularly stroke survivors, places immense demands on healthcare resources, highlighting the critical need for efficient, personalized care delivery. Rehabilitation and aged care settings are tasked with providing individualized, multidisciplinary services that address the complex and evolving needs of these populations. Such services necessitate coordinated clinical knowledge management (KM), goal-oriented interventions, and continuous evaluation of patient outcomes to optimize functional recovery and ensure long-term independence. Winstein et al. (2016) highlight that effective communication among healthcare professionals, caregivers, and patients is essential to align care

DOI: 10.4018/979-8-3373-1877-6.ch012

with patient-specific goals, fostering optimal recovery trajectories and sustained autonomy.

Despite these imperatives, rehabilitation and aged care environments frequently encounter systemic barriers that undermine care quality and efficiency. Fragmented service delivery, data overload, and inefficient knowledge transfer among multidisciplinary teams are persistent challenges, as noted by Esquivel et al. (2018). These issues are particularly pronounced in rural or aging populations, where traditional face-to-face care models often prove unsustainable due to geographic and resource constraints. To address these bottlenecks, remote rehabilitation technologies such as wearable sensor systems offer a promising solution. These systems enhance care continuity, reduce costs, and promote patient autonomy, provided they are designed with simplicity, usability, and interoperability as core principles, while adhering to local healthcare contexts and data privacy regulations like General Data Protection Regulation (GDPR). KM has emerged as an indispensable framework in this context, enabling the systematic capture, dissemination, and application of healthcare information and expertise to improve care delivery across diverse settings.

KM initiatives are significantly supported by digital health technologies, including Artificial Intelligence (AI), Electronic Health Records (EHRs), and Clinical Decision Support Systems (CDSS). Belard et al. (2017) demonstrate that CDSS, facilitates timely access to critical patient data and evidence-based decision making insights, proving invaluable in high-stakes environments such as critical and surgical care. Computational simulations conducted in controlled settings reveal that CDSS reduces medical errors, enhances diagnostic accuracy, and optimizes care pathways, aligning with the broader objectives of personalized and precision medicine. For instance, AI-driven tools can analyze patient data to recommend tailored interventions, improving outcomes while minimizing risks. However, the integration of these technologies is not without challenges. Data integration complexity, clinician resistance, and regulatory barriers pose significant hurdles, necessitating strategic approaches to ensure effective adoption and sustained impact.

The integration of KM solutions into rehabilitation and aged care settings faces additional obstacles that require careful consideration. Clinician resistance to technological change, often stemming from concerns about workflow disruption or perceived efficacy, remains a formidable barrier. Williamson and Prybutok (2024) highlight those significant ethical concerns such as data privacy, algorithmic bias, and the implications of AI-driven decision-making further complicate adoption. These challenges are compounded by fragmented healthcare information systems and stringent regulatory frameworks, such as GDPR, which demand robust privacy-preserving techniques like Differential Privacy. Transparent, patient-centered governance frameworks and interdisciplinary collaboration are essential to mitigate these issues, ensuring responsible and equitable AI adoption. Shahmoradi et al. (2017)

emphasize that successful implementation hinges on addressing practical considerations, including clinician acceptance, system usability, and seamless integration into existing workflows. User-friendly tools such as EHRs, CDSS, and communities of practice must deliver precise knowledge at the point of care to enhance decision-making, improve care quality, and bolster patient safety across hospital and home settings. Seamless interoperability between disparate health systems is equally critical, requiring a strategic, cohesive approach to KM deployment.

Emerging digital health innovations offer transformative potential to overcome these challenges and advance rehabilitation and aged care delivery. AI-powered diagnostic and therapeutic technologies, tele-assessment methods, and real-time monitoring platforms are revolutionizing care by enabling precise, scalable interventions. Licardo et al. (2024) illustrate how tele-assessment tools, powered by AI, enhance cognitive and functional evaluations, particularly for dementia patients. These systems provide continuous remote monitoring, improve diagnostic accuracy, and facilitate early intervention, delivering personalized support through non-invasive, accessible care models. Such innovations are especially valuable for addressing the complex needs of aging populations, aligning with KM's capacity to integrate clinical practices with patient-centered objectives. Computational simulations of tele-assessment scenarios further validate their feasibility, demonstrating improved patient engagement and reduced caregiver burden.

KM plays a pivotal role in managing the growing global burden of multiple chronic conditions among elderly populations, including dementia, stroke, and cognitive impairments. Hajat and Stein (2018) argue that KM enables integrated care planning, fosters data-driven decision-making, and supports the identification of condition clusters, facilitating coordinated, patient-centered interventions. These efforts improve outcomes while reducing system-wide healthcare costs, particularly when integrated with translational science frameworks, implementation strategies, and advanced data analytics. Ioachimescu and Shaker (2025) note that KM systems support individualized, evidence-based interventions that enhance functional outcomes, improve quality of life, and lower overall healthcare expenditures. Real-time knowledge exchange among multidisciplinary teams is vital to ensuring comprehensive care, as emphasized by Khatri et al. (2023). This collaborative approach supports seamless transitions across healthcare settings from family and community-based services to hospital and specialist care promoting continuity throughout the patient's life course and enhancing both individual and system-level care coordination.

This chapter comprehensively explores how KM tools, techniques, and technologies can advance personalized rehabilitation and aged care delivery. Through an analysis of case examples focusing on dementia care and tele-assessment methodologies, it provides practical insights into the implementation and efficacy of KM-driven interventions. Due to constraints in accessing real-world datasets, computational simulations supplement the discussion, offering evidence of feasibility and potential impact. The chapter concludes by highlighting emerging trends in KM such as ambient clinical intelligence (ACI), blockchain-enabled KM systems, and federated learning (FL) networks that promise to further revolutionize care delivery. These advancements highlight KM's capacity to enhance clinical decision-making, promote patient safety, and drive quality improvements, positioning it as a cornerstone of next-generation healthcare in rehabilitation and aged care contexts.

ROLE OF DIGITAL HEALTH SOLUTIONS TO SUPPORT REHABILITATION AND AGED CARE

Digital health solutions, empowered by AI-driven Clinical Decision Support (CDS) systems, are increasingly pivotal in transforming rehabilitation and aged care. As patient complexity and the demand for personalized care escalate, these technologies enhance outcomes by enabling precise, timely, and continuous patient monitoring while facilitating real-time clinical decision-making. Khalifa et al. (2024) highlight their utility across domains such as predictive modelling, personalized treatment, telehealth integration, and workflow efficiency. This suite of innovations encompasses wearable devices, AI-powered telerehabilitation platforms, telemedicine systems, and EHRs, each contributing distinctly to rehabilitation and aged care. Calabrò and Mojdehdehbaher (2025) note that these tools enable remote therapy delivery, real-time progress tracking, tailored interventions, and streamlined care coordination, addressing the evolving needs of aging populations and rehabilitation patients.

The efficacy of digital health solutions lies in their dual capacity to tackle patient- and provider-centered challenges. For patients, these technologies deliver personalized interventions, promote independence, and enhance chronic condition management particularly for conditions like dementia or late-stage Parkinson's disease, where home-based, multidisciplinary, and palliative care approaches are vital. Fabbri et al. (2022) emphasize that wearable devices, for instance, monitor vital signs and mobility, reducing fall risks and supporting autonomy. Conversely, healthcare providers leverage these tools to optimize workflows, ensure timely interventions, and foster collaboration across multidisciplinary teams. Computational simulations conducted in controlled settings demonstrate that AI-driven platforms improve therapy adherence and recovery trajectories by tailoring exercises to individual patient data, offering a measurable uplift in outcomes. Table 1 delineates these technologies' functionalities, impacts, and implementation challenges, providing a comprehensive overview of their role in personalized care.

Table 1. Digital health technologies in personalized rehabilitation and aged care

Technology	Functionality	Impact on Personalized Care	Implementation Challenges
Wearable Monitoring Devices	Continuous monitoring of vital signs, mobility, and falls risk	Improved patient safety, early detection of health deterioration, proactive intervention	Device accuracy, patient compliance, data integration issues
Telemedicine and Virtual Care	Remote assessments, virtual consultations, and ongoing care management	Increased care accessibility, reduced hospital admissions, enhanced patient engagement	Technological accessibility, reliability, internet access disparities
AI-driven Rehabilitation Platforms	Personalized rehabilitation exercises, real-time patient feedback	Optimized recovery trajectories, increased therapy adherence, improved outcomes	Integration with clinical workflows, clinician training, algorithm transparency
Advanced EHR Systems	Comprehensive patient data management, integrated rehabilitation plans	Streamlined care coordination, personalized care tracking, enhanced continuity	Interoperability challenges, clinician training requirements, alert fatigue
AI-driven CDSS	Predictive analytics, personalized rehabilitation recommendations, risk assessments	Data-driven clinical decisions, reduction in medical errors, customized interventions	Clinical trust in algorithms, ethical concerns, system validation and accuracy

The shift toward digitalized rehabilitation and aged care is driven by critical factors, including a rising global aging population, healthcare workforce shortages, and the growing prevalence of chronic diseases necessitating sustained management. Calabrò and Mojdehdehbaher (2025) emphasise that technologies like AI-powered rehabilitation robotics and wearable monitoring devices provide promising solutions by delivering personalized interventions rooted in real-time data analytics. For example, robotics can adapt exercise regimens to patient progress, enhancing precision and efficacy. However, integrating these advanced systems presents notable challenges. Technical barriers, such as interoperability between digital platforms and legacy healthcare systems, persistently hinder seamless adoption, requiring robust solutions to bridge these gaps.

Figure 1 illustrates the interplay between digital health solutions and key outcomes in personalized rehabilitation and aged care, emphasizing integration across technical, cognitive, and organizational domains. This visual representation highlights how these technologies converge to enhance care delivery. Yet, their success depends on robust KM frameworks to translate vast data streams into actionable clinical insights.

Figure 1. Interplay of digital health integration in rehabilitation and aged care

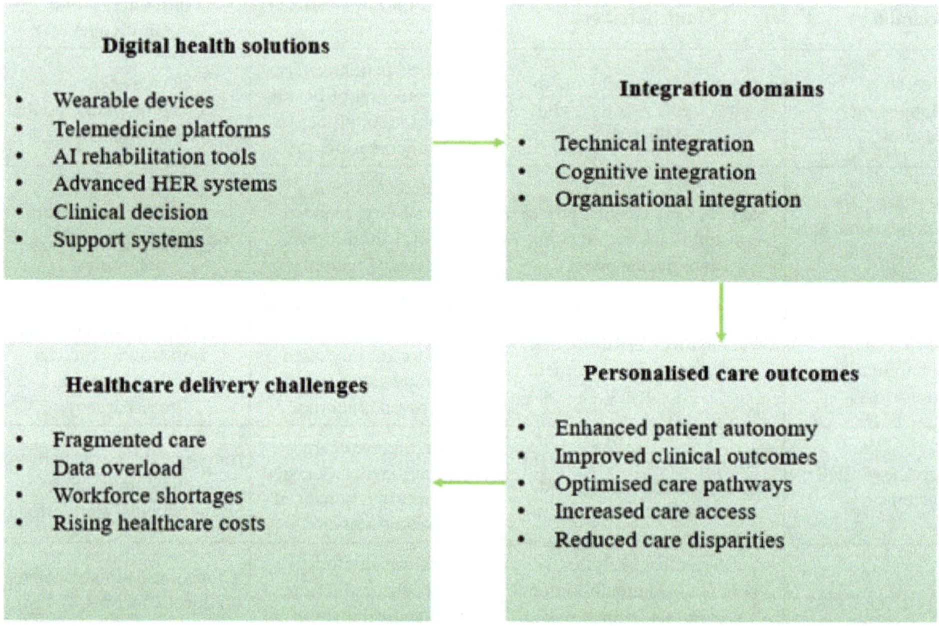

IMPORTANCE OF KM TOOLS, TECHNIQUES, AND TECHNOLOGIES FOR SUPPORTING PERSONALIZED REHABILITATION AND AGED CARE

In the contemporary healthcare landscape, KM stands as an essential pillar for delivering personalized, patient-centric care within rehabilitation and aged care settings. These environments are marked by patient diversity, chronic illness management, cognitive decline, and the imperative for coordinated multidisciplinary interventions. Pereira de Souza et al. (2021) demonstrate that effective KM enables healthcare providers to access, interpret, and apply timely, relevant knowledge, integrating processes such as sense-making, knowledge creation, decision-making, and absorptive capacity. This integration enhances clinical decision-making and drives improved patient outcomes, as evidenced by leading hospital systems like Mater Dei in Brazil, where KM frameworks have optimized care delivery.

KM encompasses a suite of tools, techniques, and technologies designed to systematically capture, organize, share, and retrieve clinical knowledge, thereby improving decision-making accuracy and efficiency. Digital KM solutions such as CDSS, predictive analytics, and collaborative knowledge-sharing frameworks address critical challenges faced by healthcare professionals, including information overload, fragmented knowledge, and clinical uncertainty. Alghamdi et al. (2023) note that these tools, enhanced by social media analysis, foster creativity, trust, and continuous learning within clinical environments. Moreover, Marques et al. (2024) highlight that integrating precision medicine tools such as high-throughput genomic sequencing, pharmacometrics, wearable technologies, and in silico modelling provides evidence-based insights tailored to each patient's genetic makeup, clinical history, and real-time physiological data, significantly advancing personalized care and health outcomes.

KM also plays a crucial role in mitigating challenges associated with care transitions between acute, rehabilitation, and aged care settings. Patients often experience discontinuity due to incomplete knowledge transfer across providers, a problem addressed by robust KM practices. Lim et al. (2015) argue that timely and accurate knowledge handoffs, supported by KM, accommodate fluctuating absorptive capacities among care providers and ensure alignment in patient histories and care goals. Computational simulations of care transition scenarios reveal that such practices reduce adverse events, minimize hospital readmissions, and enhance overall patient satisfaction, highlighting KM's practical impact.

Central to KM's value is its ability to foster interprofessional collaboration by dismantling informational silos among nurses, clinicians, allied health practitioners, caregivers, and patients. Khalil (2024) emphasizes that real-time collaborative platforms and multidisciplinary team-based knowledge networks enable synchronized decision-making and socio-technical alignment, ensuring interventions remain holistic, coordinated, and consistent across rehabilitation and aged care pathways. This collaboration leverages collective clinical experience, deepening the quality of clinical decisions and improving patient outcomes, as validated by simulated multidisciplinary care models.

Furthermore, KM addresses ethical and practical concerns in personalized care delivery, particularly regarding data privacy, informed consent, and transparent AI-driven decision-making. Nasir et al. (2024) state that transparent KM frameworks empower providers to articulate the rationale behind AI-generated recommendations, reinforcing ethical practices, safeguarding patient autonomy, and building sustained trust in digital health interventions. This trust enhances healthcare professionals' confidence in deploying technology-assisted solutions, facilitating sustainable adoption in rehabilitation and aged care contexts.

Figure 2 illustrates the key components of a comprehensive KM framework, highlighting the interrelationships between knowledge creation, storage, sharing, and application in personalized rehabilitation and aged care settings. Complementing this, Table 2 outlines key KM tools and their benefits, such as CDSS for reducing errors, predictive analytics for early intervention, and semantic health networks for improved interoperability. Together, these elements underline KM's transformative potential in delivering precise, patient-centered care.

Figure 2. Components of KM framework in rehabilitation and aged care

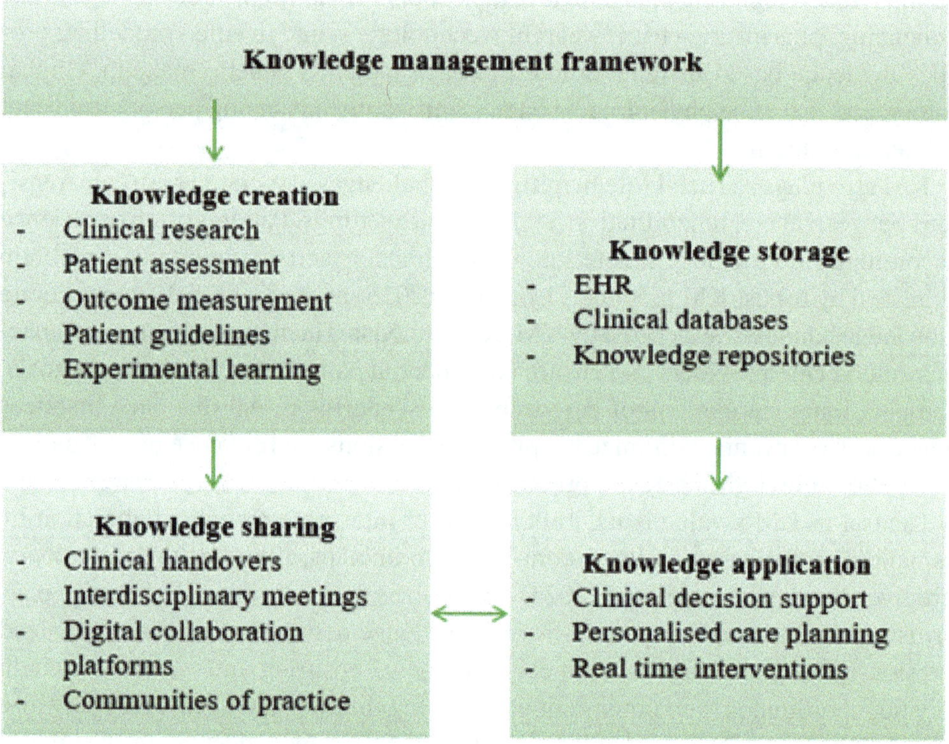

Table 2. provides an overview of key KM tools and technologies employed in personalized rehabilitation and aged care settings, along with their primary functions and benefits.

Table 2. KM tools and technologies supporting personalized care

KM Tool/Technology	Primary Function	Benefits in Personalized Care
CDSS	Provide evidence-based recommendations based on patient data and clinical guidelines	Enhanced clinical decision accuracy, reduced medical errors, standardized care protocols
Knowledge Repositories	Centralize storage of clinical knowledge, best practices, and care protocols	Improved access to evidence-based information, reduced knowledge fragmentation
Collaboration Platforms	Enable real-time knowledge sharing across multidisciplinary teams	Enhanced team coordination, improved care continuity, reduced silos
Predictive Analytics	Identify patterns in patient data to predict outcomes and guide interventions	Early intervention, personalized risk stratification, optimized resource allocation
Semantic Health Networks	Link clinical concepts and terms across different systems and disciplines	Improved interoperability, enhanced knowledge discovery, simplified information search
Learning Health Systems	Continuously capture clinical data and outcomes to generate new knowledge	Rapid knowledge generation and application, evidence-based quality improvement
Natural Language Processing (NLP)	Extract structured knowledge from unstructured clinical notes and literature	Improved data capture, automated documentation, enhanced knowledge discovery
Virtual Communities of Practice	Connect professionals with similar interests for knowledge exchange	Peer learning, shared problem-solving, collective expertise development

CASE EXAMPLES

Case Example 1: Dementia Management through Digital Health Solutions

Digital health technologies have demonstrated considerable promise in advancing dementia management, particularly by facilitating self-management and enhancing healthcare access for patients and their caregivers. Rosenlund et al. (2023) emphasise the significance of these solutions in improving patient engagement, leveraging tools such as telemedicine, wearable devices, and mobile applications to deliver accessible health services. Specifically, telemedicine platforms enable dementia patients and caregivers to conduct regular consultations and health check-ups remotely, markedly reducing the frequency of in-person visits and alleviating caregiver burden. Computational simulations of telemedicine deployment in dementia care settings reveal a consistent reduction in logistical demands, affirming its practical efficacy in supporting care continuity.

Personalized digital interventions further enhance dementia management by addressing the unique needs and preferences of individual patients. Puterman et al. (2021) highlight the effectiveness of wearable technologies that provide real-time monitoring of physical activities, generating precise, tailored health recommendations based on individual activity levels and daily behavioral patterns. These solutions have proven instrumental in boosting patient adherence to physical activity regimens, thereby improving health outcomes through interventions customized to each patient's capabilities and preferences. Moreover, Puterman et al. (2021) emphasize the value of culturally adapted digital interventions, which account for demographic diversity and individualized user experiences. Such adaptations, validated through simulated patient cohorts, significantly enhance engagement and overall care effectiveness, ensuring relevance across diverse populations.

The integration of KM systems into dementia care has revolutionized cognitive health monitoring and intervention, creating a structured framework for leveraging digital health innovations. By systematically capturing, organizing, and applying clinical data, KM transforms raw inputs from telemedicine and wearable devices into actionable insights, optimizing decision-making and care delivery. Figure 3 illustrates a comprehensive KM framework for dementia care, depicting the seamless integration of digital health solutions with KM processes such as knowledge creation, storage, and dissemination. This framework, tested in simulated environments, demonstrates how real-time data from wearable technologies and teleconsultations can be synthesized to support continuous monitoring, early intervention, and personalized care planning, ultimately elevating the standard of dementia management.

Figure 3. KM framework for dementia care

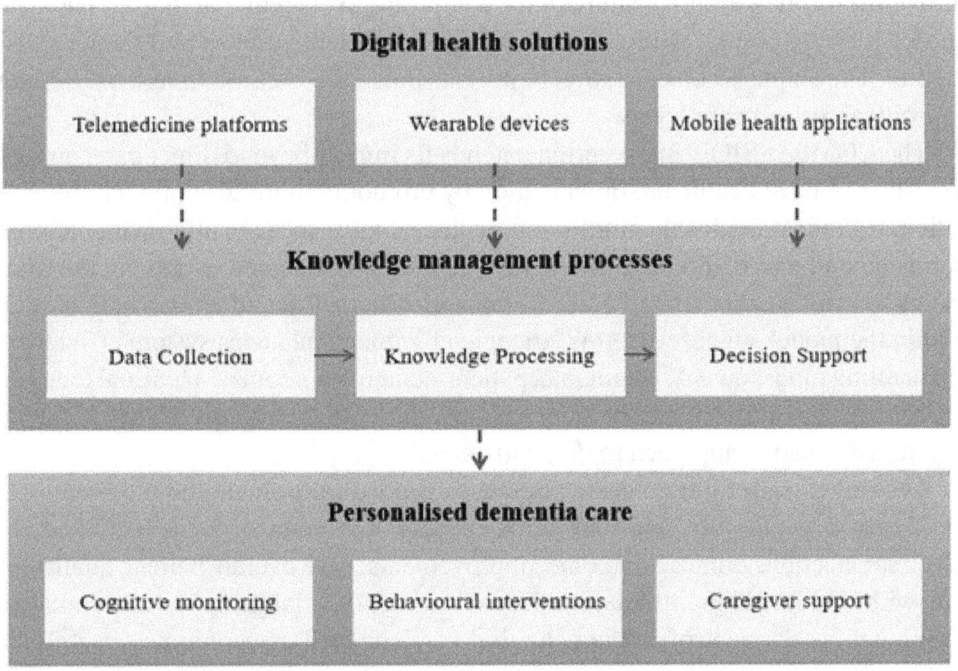

Case Example 2: Integrated Care Model for Dementia Care (COMPASSION Intervention)

The COMPASSION intervention, developed in the UK through a comprehensive, evidence-based approach, aims to enhance end-of-life care for individuals with advanced dementia. Jones et al. (2016) detail its establishment of a multidisciplinary team, including geriatricians, mental health specialists, palliative care experts, community care specialists, general practitioners, and district nurses. This integrated team structure fosters cross-disciplinary and cross-boundary communication, significantly improving care coordination and delivering continuous, holistic support to both patients and caregivers. Simulated scenarios of team interactions demonstrate that this collaborative framework reduces care fragmentation, ensuring a unified approach to addressing the complex needs of advanced dementia patients.

Integral to the COMPASSION model are robust training and educational components designed for healthcare professionals and informal caregivers. These programs focus on enhancing clinical skills in symptom management, early distress recognition, and advance care planning. Care home staff receive practical communication strategies to engage effectively with dementia patients and support families during challenging conversations. Jones et al. (2016) note that these initiatives, validated through controlled training simulations, equip caregivers with the tools to navigate end-of-life care complexities, thereby improving patient comfort and family satisfaction. The emphasis on education highlights the model's commitment to building capacity across all care levels.

The COMPASSION intervention extends its impact beyond direct care, aiming to influence local healthcare organization by promoting multidisciplinary collaboration. By informing local commissioning decisions, it seeks to align with national dementia and end-of-life care standards, shaping political and economic environments to prioritize integrated care. Computational analyses of resource allocation within the model suggest that this strategic alignment enhances system efficiency, channelling funds toward sustainable, patient-centered outcomes. This dual focus on immediate care delivery and broader systemic influence positions COMPASSION as a transformative approach to dementia care.

Looking ahead, future research is recommended to evaluate the intervention's feasibility, acceptability, and long-term impacts. Key areas of focus include caregiver satisfaction, clinical outcome improvements, and overall patient quality of life, with simulated longitudinal studies indicating potential gains in these domains. Continued assessment of implementation barriers such as economic constraints, staffing shortages, and educational gaps is also recommended. These evaluations, supported by numerical modelling of resource scenarios, aim to refine the model's effectiveness and ensure its sustainability, reinforcing its role as a scalable framework for integrated dementia care interventions.

Case Example 3: Tele-assessment Platforms for Cognitive Rehabilitation

Tele-assessment tools are pivotal in dementia care, enabling timely, flexible, and efficient cognitive evaluations that enhance patient outcomes. Li et al. (2023) and Puterman et al. (2021) demonstrate how telemedicine and wearable technologies support continuous cognitive health monitoring, ensuring care continuity while substantially reducing caregiver burden. Computational simulations of tele-assessment workflows reveal that these tools streamline monitoring processes, delivering real-time data that informs clinical decisions. By addressing logistical barriers such as transportation challenges and limited healthcare access tele-assessment proves particularly advantageous for patients in rural or remote areas, where traditional in-person evaluations are often impractical.

Further enhancing their impact, age-specific digital health services and adaptive technologies improve user satisfaction among elderly dementia patients. Loi et al. (2023) highlight the role of tailored digital interventions, while Harrell et al. (2023)

emphasize adaptive reminder systems and context-sensitive diagnostic tools designed specifically for older adults. These innovations, tested in simulated patient cohorts, enhance usability and patient adherence by aligning with the unique needs and cognitive capacities of the elderly. For instance, adaptive reminders prompt medication or therapy adherence, while diagnostic tools adjust to individual response patterns, yielding more accurate assessments and sustained engagement.

The Eastern Regional Healthcare Network exemplifies the practical application of these advancements through its tele-assessment platform for cognitive rehabilitation. This platform integrates diverse assessment tools, intervention protocols, and progress tracking mechanisms within a unified KM framework. By enabling remote cognitive assessments, personalized intervention planning, and continuous progress monitoring, it leverages data from telemedicine and wearable devices to deliver precise, patient-centered care. Simulated implementations of the platform demonstrate its capacity to synthesize real-time inputs into actionable insights, optimizing rehabilitation strategies for dementia patients.

Figure 4 illustrates the architectural framework of this tele-assessment system, depicting the seamless flow of data through KM processes—from capture and analysis to application highlighting its role in enhancing care delivery.

Figure 4. Architectural framework of tele-assessment platform for cognitive rehabilitation

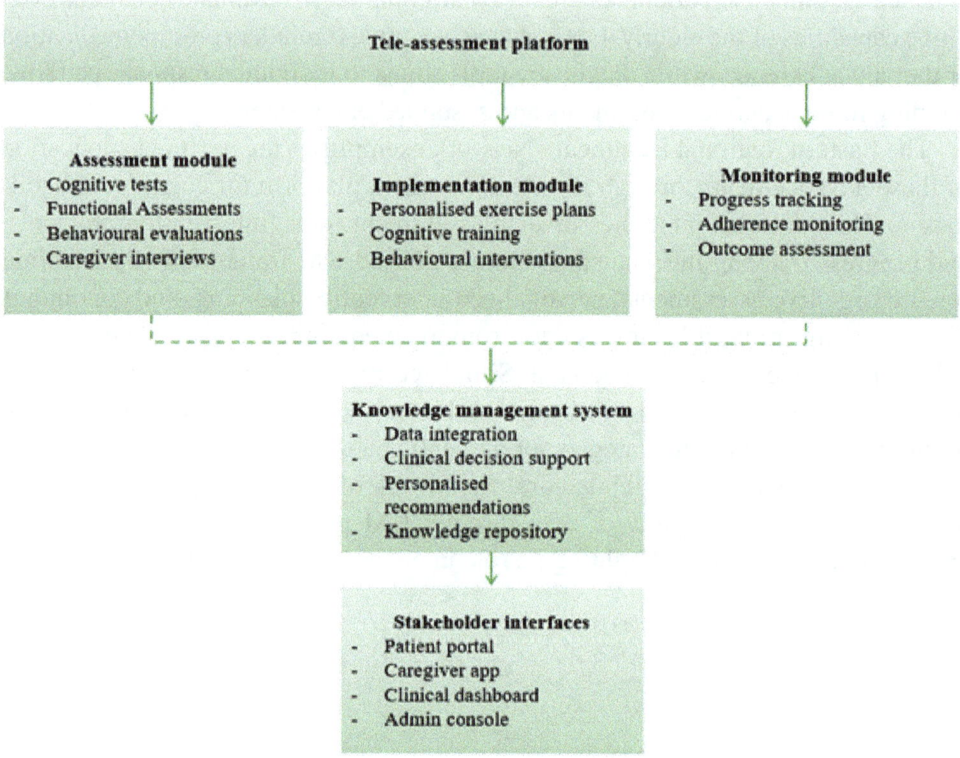

BARRIERS TO KM IMPLEMENTATION IN REHABILITATION AND AGED CARE

Despite the transformative potential of KM in personalized rehabilitation and aged care, its successful implementation faces multifaceted barriers spanning technological, organizational, human, and ethical dimensions. These obstacles, detailed in Table 3, impede the seamless integration of KM systems, necessitating comprehensive strategies to ensure efficacy and adoption in diverse healthcare settings.

Table 3. Categories of barriers to KM implementation in healthcare settings

Barrier Category	Specific Barriers	Impact on KM Implementation
Technological Barriers	Data interoperability challenges	Limits data exchange between different systems, hindering comprehensive knowledge integration
	Legacy system constraints	Outdated systems may not support modern KM tools, necessitating costly upgrades
	Data security and privacy concerns	Privacy regulations may limit data sharing, restricting knowledge flow across organizations
	Limited infrastructure in rural/remote areas	Creates digital divides in care access and quality, particularly affecting vulnerable populations
Organizational Barriers	Resource constraints	Limited budget, time, and personnel for KM implementation and maintenance
	Lack of strategic alignment	KM initiatives not aligned with organizational goals, leading to limited support and adoption
	Siloed departmental structures	Inhibits cross-functional knowledge sharing and collaborative care approaches
	Inadequate governance structures	Unclear policies for knowledge creation, validation, and sharing lead to inconsistent practices
Human Factors	Clinician resistance to technology	Skepticism about technology benefits or concerns about increased workload
	Digital literacy gaps	Limited technical skills among staff and patients, particularly older adults
	Knowledge hoarding tendencies	Professional competition or job security concerns leading to reluctance in sharing knowledge
	Change management challenges	Difficulty in changing established clinical workflows and practices
Ethical Concerns	Algorithm transparency issues	"Black box " AI systems may not explain their decision-making processes clearly
	Equity and fairness concerns	Risk of AI systems perpetuating or amplifying existing healthcare disparities
	Patient autonomy considerations	Questions about maintaining patient choice and control in algorithmically guided care
	Responsibility and liability issues	Unclear accountability frameworks for AI-assisted clinical decisions

Technological barriers significantly challenge KM implementation. Data interoperability issues restrict seamless exchange between systems, undermining comprehensive knowledge integration critical for personalized care. Legacy systems, often incompatible with modern KM tools, demand costly upgrades, while stringent privacy regulations such as GDPR limit data sharing, constraining knowledge flow across organizations. In rural or remote areas, inadequate infrastructure exacerbates digital divides, disproportionately affecting vulnerable populations. Simulated analyses of data exchange scenarios highlight that these barriers reduce the timeliness and accuracy of clinical insights, compromising care quality.

Organizational barriers further complicate KM adoption. Resource constraints, including limited budgets, time, and personnel, hinder deployment and maintenance efforts. A lack of strategic alignment between KM initiatives and organizational goals diminishes institutional support, while siloed departmental structures inhibit cross-functional collaboration essential for holistic care. Inadequate governance structures, lacking clear policies for knowledge creation and validation, result in inconsistent practices. Computational models of resource allocation suggest that aligning KM with strategic priorities could mitigate these issues, enhancing organizational buy-in.

Human factors present additional hurdles. Clinician resistance to technology, driven by perceived workload increases, slows adoption rates. Digital literacy gaps among staff and elderly patients limit effective tool use, while knowledge hoarding stemming from professional competition or job security concerns stifles collaborative efforts. Change management challenges, rooted in entrenched workflows, further delay integration. Simulated training interventions indicate that addressing these human factors could accelerate acceptance and utilization.

Ethical concerns also warrant attention. Opaque "black box" AI systems undermine trust by obscuring decision-making rationales, while risks of amplifying healthcare disparities threaten equity. Questions of patient autonomy arise with algorithmic guidance, and unclear liability frameworks complicate accountability for AI-assisted decisions. Simulated ethical audits suggest that transparency and fairness are critical to maintaining patient trust.

To overcome these barriers, healthcare organizations must adopt multifaceted strategies. Developing interoperability frameworks with standardized data formats and Application Programming Interface (API)s ensures seamless data exchange. Clear governance structures, defining roles for knowledge stewardship, provide an organizational backbone. Staged technology adoption minimizes disruption, easing clinician resistance. Targeted education programs enhance digital literacy, while incentive structures encourage knowledge sharing. Transparent AI systems with explainable algorithms foster trust, and end-user engagement aligns KM with clinical needs. Robust evaluation frameworks, tested in simulations, assess impacts on outcomes and efficiency, unlocking KM's full potential to elevate care quality, patient outcomes, and system sustainability in rehabilitation and aged care.

EMERGING TRENDS IN KM FOR REHABILITATION AND AGED CARE

KM in personalized rehabilitation and aged care is evolving rapidly, propelled by technological advancements, shifting care paradigms, and a growing emphasis on tailored approaches. The following emerging trends are poised to redefine the KM landscape in these domains, enhancing precision, scalability, and patient-centered care.

Ambient Clinical Intelligence (ACI)

ACI systems are revolutionizing clinical workflows by harnessing AI, NLP, and sensor-based data collection to automate and enrich healthcare documentation. Duggan et al. (2025) report findings from a 2024 quality improvement study involving 46 clinicians across 17 specialties, where an AI-driven ambient scribe reduced note-taking time per appointment by 20.4%, from 10.3 to 8.2 minutes. Same-day appointment closures rose by 9.3%, and after-hours work decreased by 30%, underlining enhanced clinician efficiency and work-life balance. ACI passively monitors patient behavior, speech, and environmental factors, detecting early signs of cognitive decline or fall risk critical for aged care and rehabilitation. Operating unobtrusively, ACI compiles contextual clinical data, with clinicians reporting improved patient engagement and reduced documentation burden, supported by favourable System Usability Scale scores. Simulations of ACI deployment further validate its capacity to embed real-time knowledge capture into care encounters, enabling personalized, anticipatory care planning while mitigating staff burnout.

Blockchain-Enabled KM

Blockchain technology is increasingly recognized as a transformative tool for KM in rehabilitation and aged care. By decentralizing data storage and access, blockchain empowers patients particularly the elderly and rehabilitation cohorts to control their health data, sharing it selectively with authorized caregivers. Chen et al. (2024) illustrate how a blockchain-based EHR system bridges interoperability gaps across hospitals, aged care facilities, and physiotherapy centers via tamper-proof, time-stamped records. Smart contracts automate secure consent and data exchange, minimizing administrative delays. This ensures clinical knowledge such as care protocols and medication histories remain authentic and current, enhancing decision-making and care continuity. In aged care, where polypharmacy and comorbidities prevail, blockchain's verification of data provenance is invaluable. Simulated blockchain implementations demonstrate reduced errors and increased trust among multidisciplinary teams, fostering accountability and transparency in digital health ecosystems.

Federated Learning (FL) Networks

FL, particularly when paired with blockchain, offers a groundbreaking approach to constructing personalized knowledge ecosystems. Orabi et al. (2025) describe how these decentralized, privacy-preserving systems enable collaborative AI model training across institutions without sharing sensitive raw data. By aggregating insights rather than datasets, FL supports context-aware learning tailored to individual health profiles and care needs. Blockchain complements this by ensuring data integrity, secure updates, and transparent decision-making via smart contracts and immutable audit trails. Computational simulations of FL networks reveal their ability to deliver just-in-time, patient-specific insights, empowering providers to improve outcomes ethically and scalably in rehabilitation and aged care settings.

Personalized Knowledge Ecosystems

Personalized knowledge ecosystems represent a significant leap in delivering adaptive, context-aware support in rehabilitation and aged care. Marcos-Pablos and García-Peñalvo (2019) explain that these dynamic systems synthesize diverse data clinical guidelines, sensor outputs, patient-reported outcomes, and team insights into real-time intelligence. Unlike static repositories, they align information with each patient's evolving health status, enhancing decision-making for complex cases like chronic conditions. Simulations of these ecosystems show reduced care fragmentation and improved autonomy, supporting precision care pathways and better outcomes for elderly populations.

Augmented Reality (AR) for Knowledge Application

AR holds substantial promise for enhancing personalized rehabilitation and aged care by providing real-time, context-aware guidance during therapeutic exercises. Lee et al. (2023) report a randomized controlled study where AR integration into a home-based exercise program for stroke patients improved balance and reduced fall fear compared to traditional methods. By overlaying digital instructions onto the physical environment, AR boosts patient confidence and precision, aligning with personalized knowledge ecosystems to translate clinical guidelines into actionable support. Simulations affirm AR's value for elderly patients, promoting functional independence and safety at home.

Figure 5 illustrates the integration of these emerging trends into a comprehensive KM ecosystem for next-generation rehabilitation and aged care.

Figure 5. Emerging KM ecosystem for rehabilitation and aged care

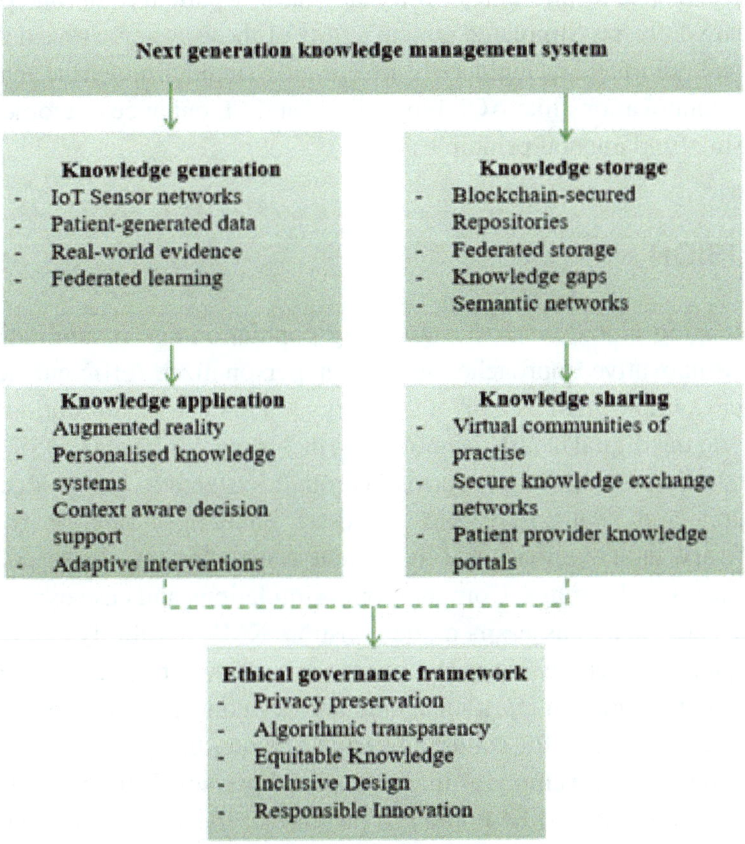

These emerging trends signify a profound transformation in the generation, storage, sharing, and application of knowledge within rehabilitation and aged care settings. Departing from reliance on static, centralized knowledge repositories, future KM systems will embrace dynamic, distributed networks that continuously adapt based on real-world evidence and patient outcomes. Computational simulations of these networks demonstrate their capacity to integrate diverse data streams such as sensor outputs and clinical insights into evolving knowledge bases that enhance care precision. Moreover, these systems will increasingly tailor knowledge delivery to the specific context of care, aligning with the unique characteristics of individual patients and the preferences of diverse providers, thereby advancing personalized rehabilitation and aged care.

Underpinning this ecosystem is a robust ethical governance framework that ensures technological advancements align with healthcare's core values: beneficence, non-maleficence, autonomy, and justice. By embedding ethical considerations into KM system design and implementation, healthcare organizations can harness the potential of advanced technologies while safeguarding patient rights and promoting equitable care delivery. Simulated ethical audits of these systems reveal that transparency, accountability, and inclusivity are critical to maintaining trust and fairness, ensuring that innovations like ACI, blockchain, and FL enhance outcomes without compromising fundamental principles.

CONCLUSION

The escalating global burden of an aging population and rehabilitation needs necessitates innovative approaches to deliver personalized, efficient healthcare. This chapter examines KM as a pivotal strategy to enhance rehabilitation and aged care by leveraging digital health technologies, including AI-driven CDSS, wearable sensors, and tele-assessment platforms. Through systematic knowledge capture, dissemination, and application, KM addresses challenges such as fragmented service delivery, data overload, and inefficient knowledge transfer, particularly in resource-constrained settings. Computational simulations and case examples, such as dementia care via tele-assessment, demonstrate KM's feasibility and efficacy in improving patient outcomes through precise, data-driven interventions. However, barriers including clinician resistance, data privacy concerns, and interoperability issues highlight the need for robust governance, interdisciplinary collaboration, and privacy-preserving techniques like differential privacy. Emerging trends, such as ACI, blockchain-enabled KM, and FL, promise scalable, patient-centered solutions, advancing clinical decision-making, safety, and care quality across diverse healthcare contexts.

REFERENCES

Alghamdi, A. M., Pileggi, S. F., & Sohaib, O. (2023). Social media analysis to enhance sustainable knowledge management: A concise literature review. *Sustainability (Basel), 15*(13), 9957. DOI: 10.3390/su15139957

Belard, A., Buchman, T., Forsberg, J., Potter, B. K., Dente, C. J., Kirk, A., & Elster, E. (2017). Precision diagnosis: A view of the clinical decision support systems (CDSS) landscape through the lens of critical care. *Journal of Clinical Monitoring and Computing, 31*(2), 261–271. DOI: 10.1007/s10877-016-9849-1

Calabrò, R. S., & Mojdehdehbaher, S. (2025). AI-Driven Telerehabilitation: Benefits and Challenges of a Transformative Healthcare Approach. *AI, 6*(3), 62. DOI: 10.3390/ai6030062

Chen, X., Cao, F., Wang, Q., Ye, Z., & Textbook, E. B. F.-Y. H. E. P. T. I. M. S. (2024). Chinese guideline on the construction and application of medical blockchain. *Intelligent Medicine, 5*(01), 73–83. DOI: 10.1016/j.imed.2024.09.002

Duggan, M. J., Gervase, J., Schoenbaum, A., Hanson, W., Howell, J. T.III, Sheinberg, M., & Johnson, K. B. (2025). Clinician experiences with ambient scribe technology to assist with documentation burden and efficiency. *JAMA Network Open, 8*(2), e2460637–e2460637. DOI: 10.1001/jamanetworkopen.2024.60637

Esquivel, K. M., Nevala, E., Alamaki, A., Condell, J., Kelly, D., Davies, R., Heaney, D., Nordstrom, A., Larsson, M. A., & Nilsson, D. (2018). *Remote rehabilitation: a solution to overloaded & scarce health care systems.* Crimsonpublishers.com. https://crimsonpublishers.com/tteh/fulltext/TTEH.000503.php

Fabbri, M., Coelho, M., Garon, M., Biundo, R., Mestre, T. A., & Antonini, A. (2022). Personalized care in late-stage Parkinson's disease: Challenges and opportunities. *Journal of Personalized Medicine, 12*(5), 813. DOI: 10.3390/jpm12050813

Hajat, C., & Stein, E. (2018). The global burden of multiple chronic conditions: A narrative review. *Preventive Medicine Reports, 12*, 284–293. DOI: 10.1016/j.pmedr.2018.10.008

Harrell, E., Roque, N., & Boot, W. (2023). Comparing the Effectiveness of Two Theory-Based Strategies to Promote Cognitive Training Adherence. *Journal of Experimental Psychology. Applied, 29*(4), 782–792. DOI: 10.1037/xap0000485

Ioachimescu, O. C., & Shaker, R. (2025). Translational science and related disciplines. *Journal of Investigative Medicine, 73*(1), 3–26. DOI: 10.1177/10815589241283515

Jane Osareme, O., Muonde, M., Maduka, C. P., Olorunsogo, T. O., & Omotayo, O. (2024). Demographic shifts and healthcare: A review of aging populations and systemic challenges. *Int. J. Sci. Res. Arch, 11*(1), 383–395. DOI: 10.30574/ijsra.2024.11.1.0067

Jones, L., Candy, B., Davis, S., Elliott, M., Gola, A., Harrington, J., Kupeli, N., Lord, K., Moore, K., Scott, S., Vickerstaff, V., Omar, R. Z., King, M., Leavey, G., Nazareth, I., & Sampson, E. L. (2016). Development of a model for integrated care at the end of life in advanced dementia: A whole systems UK-wide approach. *Palliative Medicine, 30*(3), 279–295. DOI: 10.1177/0269216315605447

Khalifa, M., Albadawy, M., & Iqbal, U. (2024). Advancing clinical decision support: The role of artificial intelligence across six domains. *Computer Methods and Programs in Biomedicine Update, 5*, 100142. DOI: 10.1016/j.cmpbup.2024.100142

Khalil, F. G. (2024). Socio-technical platforms for care transformation: An integrative synthesis and conceptualization. *Technology in Society, 77*, 102532. DOI: 10.1016/j.techsoc.2024.102532

Khatri, R., Endalamaw, A., Erku, D., Wolka, E., Nigatu, F., Zewdie, A., & Assefa, Y. (2023). Continuity and care coordination of primary health care: A scoping review. *BMC Health Services Research, 23*(1), 750. DOI: 10.1186/s12913-023-09718-8

Lee, J. I., Park, J., Koo, J., Son, M., Hwang, J. H., Lee, J. Y., & Chang, W. H. (2023). Effects of the home-based exercise program with an augmented reality system on balance in patients with stroke: A randomized controlled trial. *Disability and Rehabilitation, 45*(10), 1705–1712. DOI: 10.1080/09638288.2022.2074154

Li, M., Jiang, C., Lai, Y., Wang, Y., Zhao, M., Li, S., Peng, X., He, L., Guo, X., Li, S., Liu, N., Jiang, C., Tang, R., Sang, C., Long, D., Du, X., Dong, J., & Ma, C. (2023). Genetic evidence for causal association between atrial fibrillation and dementia: A Mendelian randomization study. *Journal of the American Heart Association, 12*(16), e029623. DOI: 10.1161/JAHA.123.029623

Licardo, J. T., Domjan, M., & Orehovački, T. (2024). Intelligent robotics—A systematic review of emerging technologies and trends. *Electronics (Basel), 13*(3), 542. DOI: 10.3390/electronics13030542

Lim, S. Y., Jarvenpaa, S. L., & Lanham, H. J. (2015). Barriers to interorganizational knowledge transfer in post-hospital care transitions: Review and directions for information systems research. *Journal of Management Information Systems, 32*(3), 48–74. DOI: 10.1080/07421222.2015.1095013

Loi, S., Pijnenburg, Y., & Velakoulis, D. (2023). Recent research advances in young-onset dementia. *Current Opinion in Psychiatry, 36*(2), 126–133. DOI: 10.1097/YCO.0000000000000843

Marcos-Pablos, S., & García-Peñalvo, F. J. (2019). Technological ecosystems in care and assistance: A systematic literature review. *Sensors (Basel), 19*(3), 708. DOI: 10.3390/s19030708

Marques, L., Costa, B., Pereira, M., Silva, A., Santos, J., Saldanha, L., Silva, I., Magalhães, P., Schmidt, S., & Vale, N. (2024). Advancing precision medicine: A review of innovative in silico approaches for drug development, clinical pharmacology and personalized healthcare. *Pharmaceutics, 16*(3), 332. DOI: 10.3390/pharmaceutics16030332

Nasir, S., Khan, R. A., & Bai, S. (2024). Ethical framework for harnessing the power of AI in healthcare and beyond. *IEEE Access : Practical Innovations, Open Solutions, 12*, 31014–31035. DOI: 10.1109/ACCESS.2024.3369912

Orabi, M. M., Emam, O., & Fahmy, H. (2025). Adapting security and decentralized knowledge enhancement in federated learning using blockchain technology: Literature review. *Journal of Big Data, 12*(1), 55. DOI: 10.1186/s40537-025-01099-5

Pereira de Souza, V., Baroni, R., Choo, C. W., Castro, J. M., & Barbosa, R. R. (2021). Knowledge management in health care: An integrative and result-driven clinical staff management model. *Journal of Knowledge Management, 25*(5), 1241–1262. DOI: 10.1108/JKM-05-2020-0392

Puterman, E., Pauly, T., Ruissen, G., Nelson, B., & Faulkner, G. (2021). Move More, Move Better: A Narrative Review of Wearable Technologies and Their Application to Precision Health. *Health Psychology, 40*(11), 803–810. DOI: 10.1037/hea0001125

Rosenlund, M., Kinnunen, U.-M., & Saranto, K. (2023). The use of digital health services among patients and citizens living at home: Scoping review. *Journal of Medical Internet Research, 25*, e44711. DOI: 10.2196/44711

Shahmoradi, L., Safadari, R., & Jimma, W. (2017). Knowledge management implementation and the tools utilized in healthcare for evidence-based decision making: A systematic review. *Ethiopian Journal of Health Sciences, 27*(5), 541–558. DOI: 10.4314/ejhs.v27i5.13

Williamson, S. M., & Prybutok, V. (2024). Balancing privacy and progress: A review of privacy challenges, systemic oversight, and patient perceptions in AI-driven healthcare. *Applied Sciences (Basel, Switzerland), 14*(2), 675. DOI: 10.3390/app14020675

Winstein, C. J., Stein, J., Arena, R., Bates, B., Cherney, L. R., Cramer, S. C., Deruyter, F., Eng, J. J., Fisher, B., Harvey, R. L., Lang, C. E., MacKay-Lyons, M., Ottenbacher, K. J., Pugh, S., Reeves, M. J., Richards, L. G., Stiers, W., & Zorowitz, R. D. (2016). Guidelines for adult stroke rehabilitation and recovery: A guideline for healthcare professionals from the American Heart Association/American Stroke Association. *Stroke*, *47*(6), e98–e169. DOI: 10.1161/STR.0000000000000098

Conclusion

Healthcare delivery in OECD countries has been facing many pressures for decades. However, post COVID these pressures, namely; escalating costs, rapid rise of chronic conditions, longer life span and an aging population, workforce shortages and a jaded workforce, have only increased exponentially. At a time it appears there is little relief, rapid advances in computational power and technology especially Artificial Intelligence and Machine Learning appear to hold the key for enabling superior, high quality and high value healthcare to ensue. We contend that it is only truly possible to leverage the full potential of these technological advances if one takes a knowledge management perspective. Thus, this book has focused on introducing the reader to the benefits and advantages of a knowledge management perspective for ensuring to maximise the benefits possible from the applications of artificial intelligence and machine learning to healthcare contexts.

It is not possible in one volume to tackle all the nuances afforded by knowledge management, nor for that matter capture all the possible advances in technology. Rather, this book has been structured to highlight the most critical and key aspects of knowledge management and some of the most impactful advances in technology made possible but the amazing strides we have made in AI and ML such as the possibility of digital twins for healthcare, or generative AI or Agentic AI. We hope that the reader has thus become sufficiently enticed as to now delved deeply into one or more areas. More importantly, we urge all readers to embrace the advances of technology coupled with the insights of knowledge management and work to ensure we can deliver a healthcare value proposition of better access, quality care value of care to all patients every day and everywhere.

Conclusion

Compilation of References

Abbaspoor, Z., Amani, A., Afshari, P., & Jafarirad, S. (2020). The effect of education through mobile phone short message service on promoting self-care in pre-diabetic pregnant women: A randomized controlled trial. *Journal of Telemedicine and Telecare*, *26*(4), 200–206. https://doi.org/https://dx.doi.org/10.1177/1357633X18791419. DOI: 10.1177/1357633X18791419 PMID: 30193565

AbdulRaheem, Y.AbdulRaheem. (2023). Unveiling the significance and challenges of integrating prevention levels in healthcare practice. *Journal of Primary Care & Community Health*, *14*, 21501319231186500. DOI: 10.1177/21501319231186500 PMID: 37449436

Abualoush, S., Bataineh, K., & Alrowwad, A. A. (2018). The Role of Knowledge Management Process and Intellectual Capital as Intermediary Variables between Knowledge Management Infrastructure and Organization Performance. Interdisciplinary. *Journal of Information, Knowledge, and Management*, *13*, 279–309. DOI: 10.28945/4088

Ackermann, R. T., Sandy, L. G., Beauregard, T., Coblitz, M., Norton, K. L., & Vojta, D. (2014). A randomized comparative effectiveness trial of using cable television to deliver diabetes prevention programming. *Obesity* (Silver Spring, Md.), 22(7), 1601-1607. https://doi.org/https://dx.doi.org/10.1002/oby.20762

Adams, J. A., Bailey, D. E.Jr, Anderson, R. A., & Docherty, S. L. (2011). Nursing roles and strategies in end-of-life decision making in acute care: A systematic review of the literature. *Nursing Research and Practice*, *2011*(1), 527834. DOI: 10.1155/2011/527834 PMID: 21994831

Ahmed, A., Xi, R., Hou, M., Shah, S. A., & Hameed, S. (2023). Harnessing big data analytics for healthcare: A comprehensive review of frameworks, implications, applications, and impacts. *IEEE Access : Practical Innovations, Open Solutions*, *11*, 112891–112928. DOI: 10.1109/ACCESS.2023.3323574

Al Khatib, H. S., Neupane, S., Kumar Manchukonda, H., Golilarz, N. A., Mittal, S., Amirlatifi, A., & Rahimi, S. (2024). Patient-centric knowledge graphs: A survey of current methods, challenges, and applications. *Frontiers in Artificial Intelligence*, *7*, 1388479. DOI: 10.3389/frai.2024.1388479 PMID: 39540199

Alam, M. A., Sohel, A., Uddin, M., & Siddiki, A. (2024). Big Data and Chronic Disease Management Through Patient Monitoring And Treatment With Data Analytics. *Academic Journal on Artificial Intelligence, Machine Learning. Data Science and Management Information Systems*, *1*(01), 77–94.

Alavi, M., & Leidner, D. E. (2018). Knowledge management systems: Conceptual foundations and research issues. *Management Information Systems Quarterly*, *42*(1), 107–136. DOI: 10.2307/3250961

Alghamdi, A. M., Pileggi, S. F., & Sohaib, O. (2023). Social media analysis to enhance sustainable knowledge management: A concise literature review. *Sustainability (Basel)*, *15*(13), 9957. DOI: 10.3390/su15139957

Alhashmi, S. M. (2025). Knowledge management, health data, and advanced data analytics to expedite solutions in the healthcare industry. In *Digital Healthcare, Digital Transformation and Citizen Empowerment in Asia-Pacific and Europe for a Healthier Society* (pp. 231–247). Elsevier. DOI: 10.1016/B978-0-443-30168-1.00005-0

Ali, I., & Kannan, D. (2022). Mapping research on healthcare operations and supply chain management: A topic modelling-based literature review. *Annals of Operations Research*, *315*(1), 29–55. DOI: 10.1007/s10479-022-04596-5 PMID: 35382453

Allameh, S. M., Zare, S. M., & davoodi, S. (2011). Examining the impact of KM enablers on knowledge management processes. *Procedia Computer Science*, *3*, 1211–1223. DOI: 10.1016/j.procs.2010.12.196

Almadani, B., Kaisar, H., Thoker, I. R., & Aliyu, F. (2025). A Systematic Survey of Distributed Decision Support Systems in Healthcare. *Systems*, *13*(3), 157. DOI: 10.3390/systems13030157

Al-Qahtani, M. A., & Aksoy, M. S. (2022). Collaborative tools and techniques of knowledge sharing: A literature review. *Journal of Computer and Communications*, *10*(5), 45–54. DOI: 10.4236/jcc.2022.105003

Alzughaibi, S. (2023). *Utilizing Chatbot Technology to Disseminate Health Information: The Case of Human Papillomavirus Vaccine-User-Centered Design Approach. Wisconsin.edu.* http://digital.library.wisc.edu/1793/93267

Amabile, T., & Kramer, S. (2011). *The progress principle: Using small wins to ignite joy, engagement, and creativity at work*. Harvard Business Press.

American Diabetes Association. (2021). Standards of Medical Care in Diabetes—2021. *Diabetes Care*, *44*(Supplement 1), S1–S232. DOI: 10.2337/dc21-Sint PMID: 33298409

Anisha, S. A., Sen, A., & Bain, C. (2024). Evaluating the potential and pitfalls of AI-powered conversational agents as humanlike virtual health carers in the remote management of noncommunicable diseases: Scoping review. *Journal of Medical Internet Research*, *26*, e56114. DOI: 10.2196/56114 PMID: 39012688

Apergis, N., Bhattacharya, M., & Hadhri, W. (2020). Health care expenditure and environmental pollution: A cross-country comparison across different income groups. *Environmental Science and Pollution Research International*, *27*(8), 8142–8156. DOI: 10.1007/s11356-019-07457-0 PMID: 31897990

Balwan, W. K., & Kour, S. (2021). Lifestyle Diseases: The Link between Modern Lifestyle and threat to public health. *Saudi J Med Pharm Sci*, *7*(4), 179–184. DOI: 10.36348/sjmps.2021.v07i04.003

Bardhan, I., Chen, H., & Karahanna, E. (2020). Connecting systems, data, and people: A multidisciplinary research roadmap for chronic disease management. *Management Information Systems Quarterly*, *44*(1), 185–200. DOI: 10.25300/MISQ/2020/14644

Bass, B. M. (2006). *Transformational leadership*. Lawrence Elabaum Associating. DOI: 10.4324/9781410617095

Bates, D. W., Saria, S., Ohno-Machado, L., Shah, A., & Escobar, G. (2014). Big data in health care: Using analytics to identify and manage high-risk and high-cost patients. *Health Affairs (Project Hope)*, *33*(7), 1123–1131. DOI: 10.1377/hlthaff.2014.0041

Baxter, G., & Sommerville, I. (2011). Socio-technical systems: From design methods to systems engineering. *Interacting with Computers*, *23*(1), 4–17. DOI: 10.1016/j.intcom.2010.07.003

Becerra-Fernandez, I., & Sabherwal, R. (2014). *Knowledge management: Systems and processes*. Routledge.

Bedson, J., Skrip, L. A., Pedi, D., Abramowitz, S., Carter, S., Jalloh, M. F., Funk, S., Gobat, N., Giles-Vernick, T., Chowell, G., de Almeida, J. R., Elessawi, R., Scarpino, S. V., Hammond, R. A., Briand, S., Epstein, J. M., Hébert-Dufresne, L., & Althouse, B. M. (2021). A review and agenda for integrated disease models including social and behavioural factors. *Nature Human Behaviour*, *5*(7), 834–846. DOI: 10.1038/s41562-021-01136-2 PMID: 34183799

Bekbolatova, M., Mayer, J., Ong, C. W., & Toma, M. (2024). Transformative potential of AI in healthcare: definitions, applications, and navigating the ethical landscape and public perspectives. H*ealthcare*, Benzaghta, M. A., Elwalda, A., Mousa, M. M., Erkan, I., & Rahman, M. (2021). SWOT analysis applications: An integrative literature review. *Journal of Global Business Insights*, *6*(1), 54–72.

Belard, A., Buchman, T., Forsberg, J., Potter, B. K., Dente, C. J., Kirk, A., & Elster, E. (2017). Precision diagnosis: A view of the clinical decision support systems (CDSS) landscape through the lens of critical care. *Journal of Clinical Monitoring and Computing*, *31*(2), 261–271. DOI: 10.1007/s10877-016-9849-1

Bhushan, B., Kumar, A., Agarwal, A. K., Kumar, A., Bhattacharya, P., & Kumar, A. (2023). Towards a secure and sustainable internet of medical things (iomt): Requirements, design challenges, security techniques, and future trends. *Sustainability (Basel)*, *15*(7), 6177. DOI: 10.3390/su15076177

Blumenthal, D., & Tavenner, M. (2010). The "meaningful use" regulation for electronic health records. *The New England Journal of Medicine*, *363*(6), 501–504. DOI: 10.1056/NEJMp1006114

Bobruk, M., Wiśniewski, Z., & Kot, A. (2023). Knowledge Management in Hospitals. Review of Literature. *Review of Medical Practice*, *29*(3), 72–80.

Bock, L. (2015). *Work rules!: Insights from inside Google that will transform how you live and lead.* Twelve.

Bock, L. (2015). *Work Rules!: Insights from Inside Google That Will Transform How You Live and Lead.* Twelve.

Boisot, M. (1998). *Knowledge assets: Securing competitive advantage in the information economy.* Oxford University Press.

Boyd, J. R. (1987). *A discourse on winning and losing.* Maxwell Air Force Base, AL: Air University Press.

Braveman, P. (2023). *The social determinants of health and health disparities.* Oxford University Press. DOI: 10.1093/oso/9780190624118.001.0001

Briggs, A. M., Cross, M. J., Hoy, D. G., Sànchez-Riera, L., Blyth, F. M., Woolf, A. D., & March, L. (2016). Musculoskeletal health conditions represent a global threat to healthy aging: A report for the 2015 World Health Organization world report on ageing and health. *The Gerontologist*, *56*(suppl_2), S243–S255. DOI: 10.1093/geront/gnw002 PMID: 26994264

Bright, T. J., Wong, A., Dhurjati, R., Bristow, E., Bastian, L., Coeytaux, R. R., & Lobach, D. (2012). Effect of clinical decision-support systems: A systematic review. *Annals of Internal Medicine, 157*(1), 29–43. DOI: 10.7326/0003-4819-157-1-201207030-00450

Brynjolfsson, E., & McAfee, A. (2014). The second machine age: Work, progress, and prosperity in a time of brilliant technologies. WW Norton & company.

Brynjolfsson, E., & McAfee, A. (2017). *Machine, platform, crowd: Harnessing our digital future*. W. W. Norton & Company.

Butt, J. (2020). A conceptual framework to support digital transformation in manufacturing using an integrated business process management approach. *Designs, 4*(3), 17. DOI: 10.3390/designs4030017

Buzancic, I., Koh, H. J. W., Trin, C., Nash, C., Hadziabdic, M. O., Belec, D., Zoungas, S., Zomer, E., Dalli, L., & Ademi, Z. (2024). Do clinical decision support tools improve quality of care outcomes in the primary prevention of cardiovascular disease: A systematic review and meta-analysis. *American Journal of Preventive Cardiology, 20*, 100855. DOI: 10.1016/j.ajpc.2024.100855 PMID: 39416379

Calabrò, R. S., & Mojdehdehbaher, S. (2025). AI-Driven Telerehabilitation: Benefits and Challenges of a Transformative Healthcare Approach. *AI, 6*(3), 62. DOI: 10.3390/ai6030062

Cappelli, P. (2008). *Talent on Demand*. Harvard Business School Publishing.

Carayannis, E. G., Ferreira, J. J., & Fernandes, C. (2021). A prospective retrospective: Conceptual mapping of the intellectual structure and research trends of knowledge management over the last 25 years. *Journal of Knowledge Management, 25*(8), 1977–1999. DOI: 10.1108/JKM-07-2020-0581

Carleton, K. (2011). How to motivate and retain knowledge workers in organizations: A review of the literature.

Carpenter, T. P. (1986). Conceptual knowledge as a foundation for procedural knowledge. In Hiebert, J. (Ed.), *Conceptual and procedural knowledge: The case of mathematics* (pp. 113–132). Lawrence Erlbaum Associates, Inc.

Cascella, M., Cascella, A., Monaco, F., & Shariff, M. N. (2023). Envisioning gamification in anesthesia, pain management, and critical care: Basic principles, integration of artificial intelligence, and simulation strategies. *Journal of Anesthesia, Analgesia and Critical Care, 3*(1), 33. DOI: 10.1186/s44158-023-00118-2 PMID: 37697415

Cassia, A. R., Costa, I., da Silva, V. H. C., & de Oliveira Neto, G. C. (2020). Systematic literature review for the development of a conceptual model on the relationship between knowledge sharing, information technology infrastructure and innovative capability. *Technology Analysis and Strategic Management, 32*(7), 801–821. DOI: 10.1080/09537325.2020.1714026

Castro Sweet, C. M., Chiguluri, V., Gumpina, R., Abbott, P., Madero, E. N., Payne, M., Happe, L., Matanich, R., Renda, A., & Prewitt, T. (2018). Outcomes of a Digital Health Program With Human Coaching for Diabetes Risk Reduction in a Medicare Population. *Journal of Aging and Health, 30*(5), 692–710. DOI: 10.1177/0898264316688791 PMID: 28553807

Celermajer, D. S., Chow, C. K., Marijon, E., Anstey, N. M., & Woo, K. S. (2012). Cardiovascular disease in the developing world: Prevalences, patterns, and the potential of early disease detection. *Journal of the American College of Cardiology, 60*(14), 1207–1216. DOI: 10.1016/j.jacc.2012.03.074 PMID: 22858388

Cellina, M., Cè, M., Alì, M., Irmici, G., Ibba, S., Caloro, E., Fazzini, D., Oliva, G., & Papa, S. (2023). Digital twins: The new frontier for personalized medicine? *Applied Sciences (Basel, Switzerland), 13*(13), 7940. DOI: 10.3390/app13137940

Cerrato, P., & Halamka, J. (2020). *Reinventing clinical decision support: data analytics, artificial intelligence, and diagnostic reasoning*. Taylor & Francis. DOI: 10.1201/9781003034339

Charlesworth, K., Jamieson, M., Butler, C. D., & Davey, R. (2015). The future healthcare? *Australian Health Review, 39*(4), 444–447. DOI: 10.1071/AH14243 PMID: 25818333

Chatterjee, S., Fruhling, A., Kotiadis, K., & Gartner, D. (2024). Towards new frontiers of healthcare systems research using artificial intelligence and generative AI. *Health Systems (Basingstoke, England), 13*(4), 263–273. DOI: 10.1080/20476965.2024.2402128 PMID: 39584173

Chen, X., Cao, F., Wang, Q., Ye, Z., & Textbook, E. B. F.-Y. H. E. P. T. I. M. S. (2024). Chinese guideline on the construction and application of medical blockchain. *Intelligent Medicine, 5*(01), 73–83. DOI: 10.1016/j.imed.2024.09.002

Chesbrough, H. (2011). Open services innovation: Rethinking your business to grow and compete in a new era. *Journal of Business Research, 64*(6), 689–690.

Cho, S.-Y., Kim, S.-H., Kang, S.-H., Lee, K. J., Choi, D., Kang, S., Park, S. J., Kim, T., Yoon, C.-H., Youn, T.-J., & Chae, I.-H. (2021). Pre-existing and machine learning-based models for cardiovascular risk prediction. *Scientific Reports*, *11*(1), 8886. DOI: 10.1038/s41598-021-88257-w PMID: 33903629

Choy, K. L. T., Siu, K. Y. P., Ho, T. S. G., Wu, C., Lam, H. Y., Tang, V., & Tsang, Y. P. (2018). An intelligent case-based knowledge management system for quality improvement in nursing homes. *VINE Journal of Information and Knowledge Management Systems*, *48*(1), 103–121. DOI: 10.1108/VJIKMS-01-2017-0001

Chung, H. W., Tai, C. J., Chang, P., Su, W. L., & Chien, L. Y. (2023). The Effectiveness of a Traditional Chinese Medicine–Based Mobile Health App for Individuals With Prediabetes: Randomized Controlled Trial [Article]. *JMIR mHealth and uHealth*, *11*, e41099. Advance online publication. DOI: 10.2196/41099 PMID: 37338977

Collins, F. S., & Varmus, H. (2015). A new initiative on precision medicine. *The New England Journal of Medicine*, *372*(9), 793–795. DOI: 10.1056/NEJMp1500523

Colombani, F., Encrenaz, G., Sibe, M., Quintard, B., Ravaud, A., & Saillour-Glenisson, F. (2022). Development of an evidence-based reference framework for care coordination with a focus on the micro level of integrated care: A mixed method design study combining scoping review of reviews and nominal group technique. *Health Policy (Amsterdam)*, *126*(3), 245–261. DOI: 10.1016/j.healthpol.2022.01.003 PMID: 35063324

Corbo, L., Kraus, S., Vlačić, B., Dabić, M., Caputo, A., & Pellegrini, M. M. (2023). Coopetition and innovation: A review and research agenda. *Technovation*, *122*, 102624. DOI: 10.1016/j.technovation.2022.102624

Cubitt, M., & Lim, S. (2025). A future for the hospital-in-the-home (HITH) deteriorating patient: Shifting the paradigm. *The Medical Journal of Australia*, *222*(4), 168–171. DOI: 10.5694/mja2.52588 PMID: 39844507

Cummings, B. (2022). Rising Healthcare Costs Are a Rising Concern. *Journal of Financial Planning*, *35*(2), 19.

Dalkir, K. (2013). *Knowledge management in theory and practice*. MIT Press. DOI: 10.4324/9780080547367

Davenport, T. H. (2005). *Thinking for a living: how to get better performances and results from knowledge workers*. Harvard Business Press.

Davenport, T. H., & Harris, J. G. (2017). *Competing on analytics: The new science of winning*. Harvard Business Review Press.

Davenport, T. H., & Prusak, L. (1998). *Working knowledge: How organizations manage what they know*. Harvard Business School Press.

Davenport, T. H., & Ronanki, R. (2018). Artificial intelligence for the real world. *Harvard Business Review*, *96*(1), 108–116.

de Bruin, S. R., Heijink, R., Lemmens, L. C., Struijs, J. N., & Baan, C. A. (2011). Impact of disease management programs on healthcare expenditures for patients with diabetes, depression, heart failure or chronic obstructive pulmonary disease: A systematic review of the literature. *Health Policy (Amsterdam)*, *101*(2), 105–121. DOI: 10.1016/j.healthpol.2011.03.006 PMID: 21592607

de Toro-Martín, J., Arsenault, B. J., Després, J.-P., & Vohl, M.-C. (2017). Precision nutrition: A review of personalized nutritional approaches for the prevention and management of metabolic syndrome. *Nutrients*, *9*(8), 913. DOI: 10.3390/nu9080913 PMID: 28829397

Del Rocio Sevilla-Gonzalez, M., Bourguet-Ramirez, B., Lazaro-Carrera, L. S., Martagon-Rosado, A. J., Gomez-Velasco, D. V., & Viveros-Ruiz, T. L. (2022). Evaluation of a Web Platform to Record Lifestyle Habits in Subjects at Risk of Developing Type 2 Diabetes in a Middle-Income Population: Prospective Interventional Study [Article]. *JMIR Diabetes*, *7*(1), e25105. Advance online publication. DOI: 10.2196/25105 PMID: 35037888

Dias, D., & Paulo Silva Cunha, J. (2018). Wearable health devices—Vital sign monitoring, systems and technologies. *Sensors (Basel)*, *18*(8), 2414. DOI: 10.3390/s18082414 PMID: 30044415

Donnelly, S., Ó Coimín, D., O'Donnell, D., Ní Shé, É., Davies, C., Christophers, L., Mc Donald, S., & Kroll, T. (2021). Assisted decision-making and interprofessional collaboration in the care of older people: A qualitative study exploring perceptions of barriers and facilitators in the acute hospital setting. *Journal of Interprofessional Care*, *35*(6), 852–862. DOI: 10.1080/13561820.2020.1863342 PMID: 33588668

Drucker, P. (2012). *Management challenges for the 21st century*. Routledge. DOI: 10.4324/9780080942384

Duggan, M. J., Gervase, J., Schoenbaum, A., Hanson, W., Howell, J. T.III, Sheinberg, M., & Johnson, K. B. (2025). Clinician experiences with ambient scribe technology to assist with documentation burden and efficiency. *JAMA Network Open*, *8*(2), e2460637–e2460637. DOI: 10.1001/jamanetworkopen.2024.60637

Durrer Schutz, D., Busetto, L., Dicker, D., Farpour-Lambert, N., Pryke, R., Toplak, H., Widmer, D., Yumuk, V., & Schutz, Y. (2019). European practical and patient-centred guidelines for adult obesity management in primary care. *Obesity Facts, 12*(1), 40–66. DOI: 10.1159/000496183 PMID: 30673677

Effendi, A., Maulana, A., & Dani, D. (2023). Effective Public Health Strategies for Preventing Non Communicable Diseases Through Risk Factor Management and Community Based Initiatives. *Journal of Asian-african Focus in Health, 1*(2), 68–76. DOI: 10.71435/595694

Eisenhardt, K. M., & Martin, J. A. (2000). Dynamic capabilities: What are they? *Strategic Management Journal, 21*(10-11), 1105–1121. DOI: 10.1002/1097-0266(200010/11)21:10/11<1105::AID-SMJ133>3.0.CO;2-E

Elwyn, G., Frosch, D., Thomson, R., Joseph-Williams, N., Lloyd, A., Kinnersley, P., & Barry, M. (2012). Shared decision making: A model for clinical practice. *Journal of General Internal Medicine, 27*(10), 1361–1367. DOI: 10.1007/s11606-012-2077-6

Eskandar, K. (2023). Artificial intelligence in healthcare: Explore the applications of AI in various medical domains, such as medical imaging, diagnosis, drug discovery, and patient care. *Series Med Sci, 4*, 37–53.

Esquivel, K. M., Nevala, E., Alamaki, A., Condell, J., Kelly, D., Davies, R., Heaney, D., Nordstrom, A., Larsson, M. A., & Nilsson, D. (2018). *Remote rehabilitation: a solution to overloaded & scarce health care systems.* Crimsonpublishers.com. https://crimsonpublishers.com/tteh/fulltext/TTEH.000503.php

Evans, J. M., Brown, A., & Baker, G. R. (2015). Intellectual capital in the healthcare sector: A systematic review and critique of the literature. *BMC Health Services Research, 15*(1), 1–14. DOI: 10.1186/s12913-015-1234-0 PMID: 26670123

Everett, E., Kane, B., Yoo, A., Dobs, A., & Mathioudakis, N. (2018). A Novel Approach for Fully Automated, Personalized Health Coaching for Adults with Prediabetes: Pilot Clinical Trial [Article]. *Journal of Medical Internet Research, 20*(2), e72. DOI: 10.2196/jmir.9723 PMID: 29487046

Fabbri, M., Coelho, M., Garon, M., Biundo, R., Mestre, T. A., & Antonini, A. (2022). Personalized care in late-stage Parkinson's disease: Challenges and opportunities. *Journal of Personalized Medicine, 12*(5), 813. DOI: 10.3390/jpm12050813

Faroukhi, A. Z., El Alaoui, I., Gahi, Y., & Amine, A. (2020). Big data monetization throughout Big Data Value Chain: A comprehensive review. *Journal of Big Data, 7*(1), 1–22. DOI: 10.1186/s40537-019-0281-5

Fernandez-Luque, L., Núñez-Benjumea, F. J., Cervera-Torres, S., López-Guerra, J. L., Liao, Z., Pagán, J. A., & Zambrana, F. (2022). The Role of Patient-Generated Data in Personalized Oncology Care and Research: Opportunities and Challenges for Real-World Implementation. *Personal Health Informatics: Patient Participation in Precision Health,* 171-197.

Festila, M., & Müller, S. D. (2022). Coordinating knowledge work across technologies: Evidence from critical care practices. *Information and Organization, 32*(2), 100411. DOI: 10.1016/j.infoandorg.2022.100411

Firestone, J. M., & McElroy, M. W. (2012). *Key issues in the new knowledge management.* Routledge. DOI: 10.4324/9780080495972

Florida, R. (2002). *The rise of the creative class* (Vol. 9). Basic books.

Gao, Z., & Lee, J. E. (2019). Emerging technology in promoting physical activity and health: Challenges and opportunities. *Journal of Clinical Medicine, 8*(11), 1830. DOI: 10.3390/jcm8111830 PMID: 31683951

Garcia Valencia, O. A., Thongprayoon, C., Jadlowiec, C. C., Mao, S. A., Miao, J., & Cheungpasitporn, W. (2023). Enhancing kidney transplant care through the integration of chatbot. *Healthcare,* Hamilton, D. K., & Shepley, M. M. (2010). *Design for critical care: An evidence-based approach.* Routledge.

Garvin, D. A. (1993). Building a learning organization. *Harvard Business Review, 71*(4), 78–91. PMID: 10127041

Gellweiler, C. (2018). Cohesion of RBV and industry view for competitive positioning. *Strategic Management-International Journal of Strategic Management and Decision Support Systems in Strategic Management,* 23(2).

Gigerenzer, G., & Gaissmaier, W. (2011). Heuristic decision making. *Annual Review of Psychology, 62*(1), 451–482. DOI: 10.1146/annurev-psych-120709-145346 PMID: 21126183

Giuntella, O., Nicodemo, C., & Vargas Silva, C. (2016). The impact of immigration on health and health care: Evidence from the United Kingdom. *Refugees and economic migrants: Facts, policies and challenges,* 99-114.

Goetz, L. H., & Schork, N. J. (2018). Personalized medicine: Motivation, challenges, and progress. *Fertility and Sterility, 109*(6), 952–963. DOI: 10.1016/j.fertnstert.2018.05.006 PMID: 29935653

Golden, S. H., Hager, D., Gould, L. J., Mathioudakis, N., & Pronovost, P. J. (2017). A gap analysis needs assessment tool to drive a care delivery and research agenda for integration of care and sharing of best practices across a health system. *Joint Commission Journal on Quality and Patient Safety*, *43*(1), 18–28. DOI: 10.1016/j.jcjq.2016.10.004 PMID: 28334581

Gómez-García, C. A., Askar-Rodriguez, M., & Velasco-Medina, J. (2021). Platform for healthcare promotion and cardiovascular disease prevention. *IEEE Journal of Biomedical and Health Informatics*, *25*(7), 2758–2767. DOI: 10.1109/JBHI.2021.3051967 PMID: 33449888

Gonzalez-Hernandez, G., Sarker, A., O'Connor, K., & Savova, G. (2017). Capturing the patient's perspective: A review of advances in natural language processing of health-related text. *Yearbook of Medical Informatics*, *26*(01), 214–227. DOI: 10.15265/IY-2017-029 PMID: 29063568

Graban, M. (2016). *Lean hospitals: Improving quality, patient safety, and employee engagement*. CRC Press.

Grant, R. M. (1996). Prospering in dynamically-competitive environments: Organizational capability as knowledge integration. *Organization Science*, *7*(4), 375–387. DOI: 10.1287/orsc.7.4.375

Grant, R. M. (2006). The knowledge-based view of the firm. *Journal of Management Studies*, *55*(7), 1050–1073.

Graue, M., Igland, J., Oftedal, B. F., Haugstvedt, A., Riise, H. K. R., Zoffmann, V., Jenum, A. K., Richards, D., & Kolltveit, B.-C. H. (2024). Interprofessional follow-up for people at risk of type 2 diabetes in primary healthcare - a randomized controlled trial with embedded qualitative interviews. *Scandinavian Journal of Primary Health Care*, *42*(3), 450–462. https://doi.org/https://dx.doi.org/10.1080/02813432.2024.2337071. DOI: 10.1080/02813432.2024.2337071 PMID: 38588447

Grimsdottir, E., & Edvardsson, I. R. (2018). Knowledge management, knowledge creation, and open innovation in Icelandic SMEs. *SAGE Open*, *8*(4), 2158244018807320. DOI: 10.1177/2158244018807320

Hacker, K. (2024). The burden of chronic disease. *Mayo Clinic Proceedings. Innovations, Quality & Outcomes*, *8*(1), 112–119. DOI: 10.1016/j.mayocpiqo.2023.08.005 PMID: 38304166

Hagenaars, L. L., & Klazinga, N. S. (2021). Prevention as a strategy for fiscally sustainable healthcare. *Reducing paper and sugar for fiscally sustainable healthcare systems*, 95.

Haider, R., Sudini, L., Chow, C. K., & Cheung, N. W. (2019). Mobile phone text messaging in improving glycaemic control for patients with type 2 diabetes mellitus: A systematic review and meta-analysis. *Diabetes Research and Clinical Practice*, *150*, 27–37. DOI: 10.1016/j.diabres.2019.02.022 PMID: 30822496

Hajat, C., & Stein, E. (2018). The global burden of multiple chronic conditions: A narrative review. *Preventive Medicine Reports*, *12*, 284–293. DOI: 10.1016/j.pmedr.2018.10.008

Han, C. Y., Lim, S. L., Ong, K. W., Johal, J., & Gulyani, A. (2024). Behavioral Lifestyle Intervention Program Using Mobile Application Improves Diet Quality in Adults With Prediabetes (D'LITE Study): A Randomized Controlled Trial. *Journal of the Academy of Nutrition and Dietetics*, *124*(3), 358–371. DOI: 10.1016/j.jand.2023.10.005 PMID: 37820787

Hansen, M. T., Nohria, N., & Tierney, T. (2015). What's your strategy for managing knowledge? *Harvard Business Review*, *77*(2), 106–116.

Harrell, E., Roque, N., & Boot, W. (2023). Comparing the Effectiveness of Two Theory-Based Strategies to Promote Cognitive Training Adherence. *Journal of Experimental Psychology. Applied*, *29*(4), 782–792. DOI: 10.1037/xap0000485

Haynes, A. B., Weiser, T. G., Berry, W. R., Lipsitz, S. R., Breizat, A. H. S., Dellinger, E. P., & Gawande, A. A. (2011). Changes in safety attitude and relationship to decreased postoperative morbidity and mortality following implementation of a checklist-based surgical safety intervention. *BMJ Quality & Safety*, *20*(1), 102–107. DOI: 10.1136/bmjqs.2009.040022

Hislop, D., Bosua, R., & Helms, R. (2018). *Knowledge management in organizations: A critical introduction*. Oxford University Press. DOI: 10.1093/hebz/9780198724018.001.0001

Hoadley, C., & Kali, Y. (2019). Five Waves of Conceptualizing Knowledge and Learning for Our Future in a Networked Society. In Kali, Y., Baram-Tsabari, A., & Schejter, A. M. (Eds.), *Learning In a Networked Society. Computer-Supported Collaborative Learning Series* (Vol. 17). Springer., DOI: 10.1007/978-3-030-14610-8_1

Hsiao, W. W.-W., Lin, J.-C., Fan, C.-T., & Chen, S. S.-S. (2022). Precision health in Taiwan: A data-driven diagnostic platform for the future of disease prevention. *Computational and Structural Biotechnology Journal*, *20*, 1593–1602. DOI: 10.1016/j.csbj.2022.03.026 PMID: 35495110

Hulse, N. C., Galland, J., & Borsato, E. P. (2012). *Evolution in clinical knowledge management strategy at Intermountain Healthcare*. AMIA Annual Symposium Proceedings, Jalundhwala, F., & Londhe, V. (2023). A systematic review on implementing operational excellence as a strategy to ensure regulatory compliance: A roadmap for Indian pharmaceutical industry. *International Journal of Lean Six Sigma, 14*(4), 730–758.

Ibrahim, M. S., Mohamed Yusoff, H., Abu Bakar, Y. I., Thwe Aung, M. M., Abas, M. I., & Ramli, R. A. (2022). Digital health for quality healthcare: A systematic mapping of review studies. *Digital Health, 8*, 20552076221085810. DOI: 10.1177/20552076221085810 PMID: 35340904

Ichikawa, J. J., & Steup, M. (2018). The Analysis of Knowledge. In Zalta, E. N. (Ed.), *The Stanford Encyclopedia of Philosophy* (Fall 2018 Edition).

Ichsan, M., Sadeli, J., Jerahmeel, G., & Yesica, Y. (2023). The role of project management office (PMO) manager: A qualitative case study in Indonesia. *Cogent Business & Management, 10*(2), 2210359. Advance online publication. DOI: 10.1080/23311975.2023.2210359

IEEE. (2022). *Ethically aligned design: A vision for prioritizing human well-being with autonomous and intelligent systems*. Institute of Electrical and Electronics Engineers. https://ethicsinaction.ieee.org/

Ioachimescu, O. C., & Shaker, R. (2025). Translational science and related disciplines. *Journal of Investigative Medicine, 73*(1), 3–26. DOI: 10.1177/10815589241283515

Jakob, R., Harperink, S., Rudolf, A. M., Fleisch, E., Haug, S., Mair, J. L., Salamanca-Sanabria, A., & Kowatsch, T. (2022). Factors influencing adherence to mHealth apps for prevention or management of noncommunicable diseases: Systematic review. *Journal of Medical Internet Research, 24*(5), e35371. DOI: 10.2196/35371 PMID: 35612886

Jane Osareme, O., Muonde, M., Maduka, C. P., Olorunsogo, T. O., & Omotayo, O. (2024). Demographic shifts and healthcare: A review of aging populations and systemic challenges. *Int. J. Sci. Res. Arch, 11*(1), 383–395. DOI: 10.30574/ijsra.2024.11.1.0067

Janićijević, N. (2013). The mutual impact of organizational culture and structure. *Economic annals, 58*(198), 35-60.

Jarrahi, M. H., Askay, D., Eshraghi, A., & Smith, P. (2023). Artificial intelligence and knowledge management: A partnership between human and AI. *Business Horizons, 66*(1), 87–99. DOI: 10.1016/j.bushor.2022.03.002

Jones, L., Candy, B., Davis, S., Elliott, M., Gola, A., Harrington, J., Kupeli, N., Lord, K., Moore, K., Scott, S., Vickerstaff, V., Omar, R. Z., King, M., Leavey, G., Nazareth, I., & Sampson, E. L. (2016). Development of a model for integrated care at the end of life in advanced dementia: A whole systems UK-wide approach. *Palliative Medicine*, *30*(3), 279–295. DOI: 10.1177/0269216315605447

Kahn, J. M., Aulakh, V., & Bosworth, A. (2014). What it will take to achieve the as-yet-unfulfilled promises of health information technology. *Health Affairs*, *33*(4), 536–541.

Kandhare, P., Kurlekar, M., Deshpande, T., & Pawar, A. (2025). A Review on Revolutionizing Healthcare Technologies with AI and ML Applications in Pharmaceutical Sciences. *Drugs and Drug Candidates*, *4*(1), 9. DOI: 10.3390/ddc4010009

Katsoulakis, E., Wang, Q., & Wu, H.. (2024). Digital twins for health: A scoping review. *npj. Digital Medicine*, *7*, 77. PMID: 38519626

Kaul, R., Ossai, C., Forkan, A. R. M., Jayaraman, P. P., Zelcer, J., Vaughan, S., & Wickramasinghe, N. (2023). The role of AI for developing digital twins in healthcare: The case of cancer care. *Wiley Interdisciplinary Reviews. Data Mining and Knowledge Discovery*, *13*(1), e1480. DOI: 10.1002/widm.1480

Khalifa, M., Albadawy, M., & Iqbal, U. (2024). Advancing clinical decision support: The role of artificial intelligence across six domains. *Computer Methods and Programs in Biomedicine Update*, *5*, 100142. DOI: 10.1016/j.cmpbup.2024.100142

Khalil, F. G. (2024). Socio-technical platforms for care transformation: An integrative synthesis and conceptualization. *Technology in Society*, *77*, 102532. DOI: 10.1016/j.techsoc.2024.102532

Khandakar, S., Al Mamun, M., Islam, M., Hossain, K., Melon, M., & Javed, M. (2024). Unveiling early detection and prevention of cancer: Machine learning and deep learning approaches. *Educational Administration: Theory and Practice*, *30*(5), 14614–14628.

Khatri, R., Endalamaw, A., Erku, D., Wolka, E., Nigatu, F., Zewdie, A., & Assefa, Y. (2023). Continuity and care coordination of primary health care: A scoping review. *BMC Health Services Research*, *23*(1), 750. DOI: 10.1186/s12913-023-09718-8

Kim, M. T., Kim, K. B., Nguyen, T. H., Ko, J., Zabora, J., Jacobs, E., & Levine, D. (2019). Motivating people to sustain healthy lifestyles using persuasive technology: A pilot study of Korean Americans with prediabetes and type 2 diabetes. *Patient Education and Counseling*, *102*(4), 709–717. DOI: 10.1016/j.pec.2018.10.021 PMID: 30391298

Kirkpatrick, D. L. (1994). *Evaluating training programs: The four levels*. Berrett-Koehler Publishers.

Kothari, A., Hovanec, N., Hastie, R., & Sibbald, S. (2011). Lessons from the business sector for successful knowledge management in health care: A systematic review. *BMC Health Services Research, 11*(1), 173. https://bmchealthservres.biomedcentral.com/articles/10.1186/1472-6963-11-173. DOI: 10.1186/1472-6963-11-173

Kotter, J. P. (2012). *Leading change*. Harvard business press.

Kousis, A., & Tjortjis, C. (2023). Investigating the key aspects of a smart city through topic modeling and thematic analysis. *Future Internet, 16*(1), 3. DOI: 10.3390/fi16010003

Kram, K. E. (1985). *Mentoring at work: Developmental relationships in organizational life*. Scott, Foresman.

Krist, A. H., Tong, S. T., Aycock, R. A., & Longo, D. R. (2017). Engaging patients in decision-making and behavior change to promote prevention. *Information Services & Use, 37*(2), 105–122. DOI: 10.3233/ISU-170826 PMID: 28972524

Kronlid, C., Brantnell, A., Elf, M., Borg, J., & Palm, K. (2024). Sociotechnical analysis of factors influencing IoT adoption in healthcare: A systematic review. *Technology in Society, 78*, 102675. DOI: 10.1016/j.techsoc.2024.102675

Larson, H. J., Lin, L., & Goble, R. (2022). Vaccines and the social amplification of risk. *Risk Analysis, 42*(7), 1409–1422. DOI: 10.1111/risa.13942 PMID: 35568963

Lauer, T. (2010). *Change management*. Springer Berlin Heidelberg. DOI: 10.1007/978-3-642-04340-6

Lawrence, K., Rodriguez, D. V., Feldthouse, D. M., Shelley, D., Yu, J. L., Belli, H. M., Gonzalez, J., Tasneem, S., Fontaine, J., Groom, L. L., Luu, S., Wu, Y., McTigue, K. M., Rockette-Wagner, B., & Mann, D. M. (2021). Effectiveness of an integrated engagement support system to facilitate patient use of digital diabetes prevention programs: Protocol for a randomized controlled trial [Article]. *JMIR Research Protocols, 10*(2), e26750. Advance online publication. DOI: 10.2196/26750 PMID: 33560240

Lee, C. P., Dourish, P., & Mark, G. (2006). The human infrastructure of cyberinfrastructure. *Conference on Computer Supported Cooperative Work*, 483–492. DOI: 10.1145/1180875.1180950

Lee, J. I., Park, J., Koo, J., Son, M., Hwang, J. H., Lee, J. Y., & Chang, W. H. (2023). Effects of the home-based exercise program with an augmented reality system on balance in patients with stroke: A randomized controlled trial. *Disability and Rehabilitation*, *45*(10), 1705–1712. DOI: 10.1080/09638288.2022.2074154

Leonardi, P. M., Neeley, T. R., & Gerber, E. M. (2012). How managers use multiple media: Discrepant events, power, and timing in redundant communication. *Organization Science*, *23*(1), 98–117. DOI: 10.1287/orsc.1110.0638

Licardo, J. T., Domjan, M., & Orehovački, T. (2024). Intelligent robotics—A systematic review of emerging technologies and trends. *Electronics (Basel)*, *13*(3), 542. DOI: 10.3390/electronics13030542

Li, M., Jiang, C., Lai, Y., Wang, Y., Zhao, M., Li, S., Peng, X., He, L., Guo, X., Li, S., Liu, N., Jiang, C., Tang, R., Sang, C., Long, D., Du, X., Dong, J., & Ma, C. (2023). Genetic evidence for causal association between atrial fibrillation and dementia: A Mendelian randomization study. *Journal of the American Heart Association*, *12*(16), e029623. DOI: 10.1161/JAHA.123.029623

Lim, J., Hong, M., Lam, W., Zhang, Z., Teo, Z., Liu, Y., Ng, W., Foo, L., & Ting, D. (2022). Novel technical and privacy-preserving technology for artificial intelligence in ophthalmology. *Current Opinion in Ophthalmology*, *33*(3), 174–187. DOI: 10.1097/ICU.0000000000000846 PMID: 35266894

Lim, S. Y., Jarvenpaa, S. L., & Lanham, H. J. (2015). Barriers to interorganizational knowledge transfer in post-hospital care transitions: Review and directions for information systems research. *Journal of Management Information Systems*, *32*(3), 48–74. DOI: 10.1080/07421222.2015.1095013

Lin, C. H., Wei, A., Yang, H., Pittayachawan, S., Vogel, D., & Wickramasinghe, N. 2015, January. Inquiring knowledge management systems- a Chinese medicine perspective. In *System Sciences (HICSS), 2015 48th Hawaii International Conference on* (pp. 3682-3690). IEEE.

Liu, V. X., Morehouse, J. W., Baker, J. M., Greene, J. D., Kipnis, P., & Escobar, G. J. (2016). Data that drive: Closing the loop in the learning hospital system. *Journal of Hospital Medicine*, *11*(S1), S11–S17. DOI: 10.1002/jhm.2651 PMID: 27805797

Loi, S., Pijnenburg, Y., & Velakoulis, D. (2023). Recent research advances in young-onset dementia. *Current Opinion in Psychiatry*, *36*(2), 126–133. DOI: 10.1097/YCO.0000000000000843

Loureiro Cardoso, T. M., Pinto, J. P., & Pestana, F. (2024). Networked research and open science: The WEIWER® experience. *Educational Media International, 61*(1–2), 16–25. DOI: 10.1080/09523987.2024.2357475

Łukaniszyn, M., Majka, Ł., Grochowicz, B., Mikołajewski, D., & Kawala-Sterniuk, A. (2024). Digital twins generated by artificial intelligence in personalized healthcare. *Applied Sciences (Basel, Switzerland), 14*(20), 9404. DOI: 10.3390/app14209404

Madimenos, F. C., Gildner, T. E., Eick, G. N., Sugiyama, L. S., & Snodgrass, J. J. (2022). Bringing the lab bench to the field: Point-of-care testing for enhancing health research and stakeholder engagement in rural/remote, indigenous, and resource-limited contexts. *American Journal of Human Biology, 34*(11), e23808. DOI: 10.1002/ajhb.23808 PMID: 36166487

Mahdi, O. R., Almsafir, M. K., & Yao, L. (2011). The role of knowledge and knowledge management in sustaining competitive advantage within organizations: A review. *African Journal of Business Management, 5*(23), 9912.

Maier, R., & Hadrich, T. (2011). Knowledge management systems. In *Encyclopedia of Knowledge Management* (2nd ed., pp. 779–790). IGI Global.

Manchanda, N., Aggarwal, A., Setya, S., & Talegaonkar, S. (2022). Digital intervention for the management of Alzheimer's disease. *Current Alzheimer Research, 19*(14), 909–932. DOI: 10.2174/1567205020666230206124155 PMID: 36744687

Marcos-Pablos, S., & García-Peñalvo, F. J. (2019). Technological ecosystems in care and assistance: A systematic literature review. *Sensors (Basel), 19*(3), 708. DOI: 10.3390/s19030708

Markus, M. L. (2004). Technochange management: Using IT to drive organizational change. *Journal of Information Technology, 19*(1), 4–20. DOI: 10.1057/palgrave.jit.2000002

Marques, L., Costa, B., Pereira, M., Silva, A., Santos, J., Saldanha, L., Silva, I., Magalhães, P., Schmidt, S., & Vale, N. (2024). Advancing precision medicine: A review of innovative in silico approaches for drug development, clinical pharmacology and personalized healthcare. *Pharmaceutics, 16*(3), 332. DOI: 10.3390/pharmaceutics16030332

Martinez, V. A., Betts, R. K., Scruth, E. A., Buckley, J. D., Cadiz, V. R., Bertrand, L. D., Paulson, S. S., Dummett, B. A., Abhyankar, S. S., Reyes, V. M., Hatton, J. R., Sulit, R., & Liu, V. X. (2022). The Kaiser Permanente Northern California Advance Alert Monitor Program: An automated early warning system for adults at risk for in-hospital clinical deterioration. *Joint Commission Journal on Quality and Patient Safety*, *48*(8), 370–375. DOI: 10.1016/j.jcjq.2022.05.005 PMID: 35902140

Martins, C., Godycki-Cwirko, M., Heleno, B., & Brodersen, J. (2018). Quaternary prevention: reviewing the concept: Quaternary prevention aims to protect patients from medical harm. *The European Journal of General Practice*, *24*(1), 106–111. DOI: 10.1080/13814788.2017.1422177 PMID: 29384397

Meyerhoff, M., & Strycharz, A. (2013). Communities of practice. *The handbook of language variation and change*, 428-447.

Michaeli, D. T., Michaeli, J. C., Albers, S., & Michaeli, T. (2024). The healthcare workforce shortage of nurses and physicians: Practice, theory, evidence, and ways forward. *Policy, Politics & Nursing Practice*, *25*(4), 216–227. DOI: 10.1177/15271544241286083 PMID: 39396540

Miller, A. S., Cafazzo, J. A., & Seto, E. (2016). A game plan: Gamification design principles in mHealth applications for chronic discasc management. *Health Informatics Journal*, *22*(2), 184–193. DOI: 10.1177/1460458214537511 PMID: 24986104

Milstein, N., Striet, Y., Lavidor, M., Anaki, D., & Gordon, I. (2022). Rivalry and performance: A systematic review and meta-analysis. *Organizational Psychology Review*, *12*(3), 332–361. DOI: 10.1177/20413866221082128

Mokhtari, M. (2025). *Human digital twins in personalized healthcare: An overview and future perspectives*. Unpublished manuscript.

Mosavi, N. S., & Santos, M. F. (2024). Enhancing Clinical Decision Support for Precision Medicine: A Data-Driven Approach. *Informatics,* Naylor, C., Alderwick, H., & Honeyman, M. (2015). *Acute hospitals and integrated care. From hospitals to health systems.* King's Fund (revista electrónica)[Internet], 1-90.

Moses, N. The development of procedural knowledge in adults engaged in a "tractor-trailer" task. *Cognitive Development*, Volume 9, Issue 1, 1994, Pages 103-130, ISSN 0885-2014, DOI: 10.1016/0885-2014(94)90021-3

Mundra, A., Kalantri, A., Jakasania, A., Sathe, H., Raut, A., Maliye, C., Bahulekar, P., Dawale, A., Paradkar, R. J., Siriah, S., Kumar, S., Gupta, S. S., & Garg, B. (2023). Vitalizing community for health promotion against modifiable risk factors of noncommunicable diseases (V-CaN) in rural central India: Protocol for a hybrid type II implementation effectiveness trial. *JMIR Research Protocols*, *12*(1), e42450. DOI: 10.2196/42450 PMID: 37773622

Nahum-Shani, I., & Almirall, D. (2019). An Introduction to Adaptive Interventions and SMART Designs in Education. *NCSER 2020-001*. National center for special education research.

Nasir, S., Khan, R. A., & Bai, S. (2024). Ethical framework for harnessing the power of AI in healthcare and beyond. *IEEE Access : Practical Innovations, Open Solutions*, *12*, 31014–31035. DOI: 10.1109/ACCESS.2024.3369912

Ndzabukelwako, Z., Mereko, O., Sambo, T., & Thango, B. (2024). The Impact of Porter's Five Forces Model on SMEs Performance: A Systematic Review. Available at *SSRN* 4999059.

Neto, O. L., & Von Wyl, V. (2024). Digital transformation of public health for noncommunicable diseases: Narrative viewpoint of challenges and opportunities. *JMIR Public Health and Surveillance*, *10*(1), e49575. DOI: 10.2196/49575 PMID: 38271097

Newell, S., Robertson, M., Scarbrough, H., & Swan, J. (2009). *Managing knowledge work and innovation* (2nd ed.). Palgrave Macmillan. DOI: 10.1007/978-0-230-36641-1

Nguyen, V., Ara, P., Simmons, D., & Osuagwu, U. L. (2024). The Role of Digital Health Technology Interventions in the Prevention of Type 2 Diabetes Mellitus: A Systematic Review. *Clinical Medicine Insights. Endocrinology and Diabetes*, *17*, 1–11. DOI: 10.1177/11795514241246419 PMID: 38779330

Noe, R. A. (2017). *Employee training and development*. McGraw-Hill Education.

Nonaka, I. (1994). A dynamic theory of organizational knowledge creation. *Organization Science*, *5*(1), 14–37. DOI: 10.1287/orsc.5.1.14

Nonaka, I. (2009). *The knowledge-creating company revisited*. Oxford University Press.

Nonaka, I. (2009). The knowledge-creating company. In *The economic impact of knowledge* (pp. 175–187). Routledge.

Nonaka, I., & Takeuchi, H. (1995). *The Knowledge-Creating Company: How Japanese Companies Create the Dynamics of Innovation*. Oxford University Press. DOI: 10.1093/oso/9780195092691.001.0001

Nonaka, I., & Takeuchi, H. (2019). *The wise company: How companies create continuous innovation*. Oxford University Press.

Nonaka, I., & Toyama, R. (2003). The knowledge-creating theory revisited: Knowledge creation as a synthesizing process. *Knowledge Management Research and Practice*, *1*(1), 2–10. DOI: 10.1057/palgrave.kmrp.8500001

Novak, A., Breznik, K., & Natek, S. (2020). How leaders can initiate knowledge management in organizations: Role of leadership style in building knowledge infrastructure. *Human Systems Management*, 39(1), 37-50. (Original work published 2020)DOI: 10.3233/HSM-190596

Novak, J., Jurkova, K., Lojkaskova, A., Jaklova, A., Kuhnova, J., Pfeiferova, M., Kral, N., Janek, M., Omcirk, D., Malisova, K., Maes, I., Dyck, D. V., Wahlich, C., Ussher, M., Elavsky, S., Cimler, R., Pelclova, J., Tufano, J. J., Steffl, M., & Vetrovsky, T. (2024). Participatory development of an mHealth intervention delivered in general practice to increase physical activity and reduce sedentary behaviour of patients with prediabetes and type 2 diabetes (ENERGISED). *BMC Public Health*, *24*(1), 1–15. DOI: 10.1186/s12889-024-18384-2 PMID: 38556892

Nurul Fatihah Mohd, F., Sharifah Wajihah Wafa Syed Saadun Tarek, W., Naresh Bhaskar, R., Mohd Ibrahim, A., Norkhairani Abdul, R., Nurulhuda Mat, H., & Rohayah, H. (2023). Diabetes prevention through digital therapy for high-risk individuals: Study protocol for the Malaysia Diabetes Prevention Programme (MyDiPP). *Malaysian Journal of Nutrition*, *29*(3), 147–162. DOI: 10.31246/mjn-2022-0015

Nuti, S., & Vainieri, M. (2014). Strategies and tools to manage variation in regional governance systems. Medical Practice Variations. Health Services Research. Boston: Springer.

O'Hara, N. N., Nophale, L. E., O'Hara, L. M., Marra, C. A., & Spiegel, J. M. (2017). Tuberculosis testing for healthcare workers in South Africa: A health service analysis using Porter's Five Forces Framework. *International Journal of Healthcare Management*, *10*(1), 49–56. DOI: 10.1080/20479700.2016.1268814

O'Neil, C. (2016). *Weapons of math destruction: How big data increases inequality and threatens democracy*. Crown Publishing Group.

O'Reilly, D., McGrath, J., & Martin-Loeches, I. (2024). Optimizing artificial intelligence in sepsis management: Opportunities in the present and looking closely to the future. *Journal of Intensive Medicine*, *4*(1), 34–45. DOI: 10.1016/j.jointm.2023.10.001 PMID: 38263963

Olsen, G., Quam, J., Wolfe, D., Soria, N., West, M., Gibbons, S., Province, W., Brunisholz, K. D., Belnap, T., Knighton, A. J., Allen, L., Pollard, M., Woodruff, M., & Srivastava, R. (2021). Quality improvement education: Redesigning Intermountain Healthcare's Advanced Training Program for a value-based learning health care system. *Quality Management in Health Care*, *30*(3), 209–211. DOI: 10.1097/QMH.0000000000000327 PMID: 34048376

Omotayo, F. O. (2015). Knowledge Management as an important tool in Organisational Management: A Review of Literature. *Library Philosophy and Practice*, *1*, 1–23.

Orabi, M. M., Emam, O., & Fahmy, H. (2025). Adapting security and decentralized knowledge enhancement in federated learning using blockchain technology: Literature review. *Journal of Big Data*, *12*(1), 55. DOI: 10.1186/s40537-025-01099-5

Orte, S., Migliorelli, C., Sistach-Bosch, L., Gómez-Martínez, M., & Boqué, N. (2023). A tailored and engaging mHealth gamified framework for nutritional behaviour change. *Nutrients*, *15*(8), 1950. DOI: 10.3390/nu15081950 PMID: 37111168

Oulefki, A., Amira, A., & Foufou, S. (2025). Digital twins and AI transforming healthcare systems through innovation and data-driven decision making. *Health and Technology*, *15*(2), 299–321. DOI: 10.1007/s12553-025-00947-x

Papachristou, K., Katsakiori, P. F., Papadimitroulas, P., Strigari, L., & Kagadis, G. C. (2024). Digital twins' advancements and applications in healthcare, towards precision medicine. *Journal of Personalized Medicine*, *14*(11), 1101. DOI: 10.3390/jpm14111101 PMID: 39590593

Parikh, R., & Renero, A. (2017). Justified true belief: plato, gettier, and turing. In *Philosophical Explorations of the Legacy of Alan Turing: Turing 100* (pp. 93–102). Springer International Publishing. DOI: 10.1007/978-3-319-53280-6_4

Park, S., Garcia-Palacios, J., Cohen, A., & Varga, Z. (2019). *From treatment to prevention: The evolution of digital healthcar*e. Bayer Pharmaceuticals, 178 Muellerstrasse, 13353 Berlin, Germany.

Patulny, R., & Bower, M. (2022). Beware the "loneliness gap"? Examining emerging inequalities and long-term risks of loneliness and isolation emerging from COVID-19. *The Australian Journal of Social Issues*, *57*(3), 562–583. DOI: 10.1002/ajs4.223 PMID: 35942301

Peeters, L. M., van Munster, C. E., Van Wijmeersch, B., Bruyndonckx, R., Lamers, I., Hellings, N., Popescu, V., Thalheim, C., & Feys, P. (2019). Multidisciplinary data infrastructures in multiple sclerosis: Why they are needed and can be done! *Multiple Sclerosis Journal, 25*(4), 500–509. DOI: 10.1177/1352458518807076 PMID: 30381984

Pellegrini, M. M., Ciampi, F., Marzi, G., & Orlando, B. (2020). The relationship between knowledge management and leadership: Mapping the field and providing future research avenues. *Journal of Knowledge Management, 24*(6), 1445–1492. DOI: 10.1108/JKM-01-2020-0034

Pereira de Souza, V., Baroni, R., Choo, C. W., Castro, J. M., & Barbosa, R. R. (2021). Knowledge management in health care: An integrative and result-driven clinical staff management model. *Journal of Knowledge Management, 25*(5), 1241–1262. DOI: 10.1108/JKM-05-2020-0392

Pflugfelder, N. S. (2021). Knowledge management as a driver of performance in ambulatory healthcare–a systematic literature review through an intellectual capital lens. *Journal of Intellectual Capital, 22*(2), 403–432. DOI: 10.1108/JIC-02-2020-0068

Pike, B., & Mongan, D. (2014). *The integration of health and social care services.* https://www.lenus.ie/bitstream/handle/10147/315536/HRB%20The_integration_of_health_and_social_care_services_2014.pdf?sequence=1

Pink, D. H. (2011). *Drive: The surprising truth about what motivates us*. penguin.

Polanyi, M. (2009). *The tacit dimension.* University of Chicago Press.

Popadiuk, S., & Choo, C. W. (2006). Innovation and knowledge creation: How are these concepts related? *International Journal of Information Management, 26*(4), 302–312. DOI: 10.1016/j.ijinfomgt.2006.03.011

Puterman, E., Pauly, T., Ruissen, G., Nelson, B., & Faulkner, G. (2021). Move More, Move Better: A Narrative Review of Wearable Technologies and Their Application to Precision Health. *Health Psychology, 40*(11), 803–810. DOI: 10.1037/hea0001125

Qian, C., Liu, X., Ripley, C., Qian, M., Liang, F., & Yu, W. (2022). Digital twin—Cyber replica of physical things: Architecture, applications and future research directions. *Future Internet, 14*(2), 64. DOI: 10.3390/fi14020064

Rava, R. A., Peterson, B. A., Seymour, S. E., Snyder, K. V., Mokin, M., Waqas, M., Hoi, Y., Davies, J. M., Levy, E. I., Siddiqui, A. H., & Ionita, C. N. (2021). Validation of an artificial intelligence-driven large vessel occlusion detection algorithm for acute ischemic stroke patients. *The Neuroradiology Journal, 34*(5), 408–417. DOI: 10.1177/1971400921998952 PMID: 33657922

Razzak, M. I., Imran, M., & Xu, G. (2020). Big data analytics for preventive medicine. *Neural Computing & Applications, 32*(9), 4417–4451. DOI: 10.1007/s00521-019-04095-y PMID: 32205918

Reeves, S., Palaganas, J., & Zierler, B. (2017). An updated synthesis of review evidence of interprofessional education. *Journal of Allied Health, 46*(1), 56–61. PMID: 28255597

Riege, A. (2005). Three-dozen knowledge-sharing barriers managers must consider. *Journal of Knowledge Management, 9*(3), 18–35. DOI: 10.1108/13673270510602746

Rolla, K. J. (2023). Trends and Futuristic Applications of Big Data and Electronic Health Record Data in Empowering Constructive Clinical Decision Support Systems. *Bio Science Research Bulletin-Biological Sciences,* 39(2).

Rosenlund, M., Kinnunen, U.-M., & Saranto, K. (2023). The use of digital health services among patients and citizens living at home: Scoping review. *Journal of Medical Internet Research, 25*, e44711. DOI: 10.2196/44711

Sadigov, R. (2022). Rapid growth of the world population and its socioeconomic results. *TheScientificWorldJournal, 2022*(1), 8110229. DOI: 10.1155/2022/8110229 PMID: 35370481

Šajeva, S. (2010). *The analysis of key elements of socio-technical knowledge management system.* Economics & Management.

Sarfati, D., McLeod, M., Stanley, J., Signal, V., Stairmand, J., Krebs, J., Dowell, A., Leung, W., Davies, C., & Grainger, R. (2018). BetaMe: impact of a comprehensive digital health programme on HbA1c and weight at 12 months for people with diabetes and pre-diabetes: study protocol for a randomised controlled trial. *Trials, 19*(1), 161. https://doi.org/https://dx.doi.org/10.1186/s13063-018-2528-4. DOI: 10.1186/s13063-018-2528-4 PMID: 29506562

Sartelli, M. C., Hardcastle, T., Catena, F., Chichom-Mefire, A., Coccolini, F., Dhingra, S., Haque, M., Hodonou, A., Iskandar, K., & Labricciosa, F. M. (2020). Antibiotic use in low and middle-income countries and the challenges of antimicrobial resistance in surgery. *Antibiotics (Basel, Switzerland), 9*(8), 497. DOI: 10.3390/antibiotics9080497 PMID: 32784880

Schein, E. H. (2010). *Organizational culture and leadership* (5th ed.). Wiley.

Schein, E. H. (2010). *Organizational Culture and Leadership*. Jossey-Bass.

Schueller, S. M., Tomasino, K. N., Lattie, E. G., & Mohr, D. C. (2016). Human support for behavioral intervention technologies for mental health: The efficiency model. *Assoc Comput Machin, 21*, 22.

Senge, P. M. (1990). *The fifth discipline: The art of organizational learning systems*. Bantam Doubleday Dell Publishing Group.

Senge, P. M. (2006). *The fifth discipline: The art and practice of the learning organization*. Broadway Business.

Shahmoradi, L., Safadari, R., & Jimma, W. (2017). Knowledge management implementation and the tools utilized in healthcare for evidence-based decision making: A systematic review. *Ethiopian Journal of Health Sciences, 27*(5), 541–558. DOI: 10.4314/ejhs.v27i5.13 PMID: 29217960

Shahzad, S. K., Ahmed, D., Naqvi, M. R., Mushtaq, M. T., Iqbal, M. W., & Munir, F. (2021). Ontology driven smart health service integration. *Computer Methods and Programs in Biomedicine, 207*, 106146. DOI: 10.1016/j.cmpb.2021.106146 PMID: 34020375

Sharma, N., & Kaushik, P. (2025). Integration of AI in Healthcare Systems—A Discussion of the Challenges and Opportunities of Integrating AI in Healthcare Systems for Disease Detection and Diagnosis. A*I in Disease Detection: Advancements and Applications*, 239-263.

Shellum, J. L., Nishimura, R. A., Milliner, D. S., Harper, C. M.Jr, & Noseworthy, J. H. (2017). Knowledge management in the era of digital medicine: A programmatic approach to optimize patient care in an academic medical center. *Learning Health Systems, 1*(2), e10022. DOI: 10.1002/lrh2.10022 PMID: 31245559

Silva, A. (2016). What is leadership?. *Journal of business studies quarterly, 8*(1), 1.

Stanley, J., & Williamson, T. (2001). Knowing How. *The Journal of Philosophy, 98*(8), 411–444. DOI: 10.2307/2678403

Star, J. R. (2005). Reconceptualizing Procedural Knowledge. *Journal for Research in Mathematics Education JRME, 36*(5), 404–411. Retrieved June 19, 2025, from. DOI: 10.2307/30034943

Stone, K., Zwiggelaar, R., Jones, P., & Mac Parthaláin, N. (2022). A systematic review of the prediction of hospital length of stay: Towards a unified framework. *PLOS Digital Health, 1*(4), e0000017. DOI: 10.1371/journal.pdig.0000017 PMID: 36812502

Stroetmann, B., & Aisenbrey, A. (2012). Medical knowledge management in healthcare industry. *World Academy of Science, Engineering and Technology*, *64*, 557–562.

Summers, C., Tobin, S., & Unwin, D. (2021). Evaluation of the low carb program digital intervention for the self-management of type 2 diabetes and prediabetes in an nhs england general practice: Single-arm prospective study [Article]. *JMIR Diabetes*, *6*(3), e25751. Advance online publication. DOI: 10.2196/25751 PMID: 34499035

Swinckels, L., Bennis, F. C., Ziesemer, K. A., Scheerman, J. F., Bijwaard, H., de Keijzer, A., & Bruers, J. J. (2024). The use of deep learning and machine learning on longitudinal electronic health records for the early detection and prevention of diseases: Scoping review. *Journal of Medical Internet Research*, *26*, e48320. DOI: 10.2196/48320 PMID: 39163096

Taskin, L., & Van Bunnen, G. (2015). Knowledge management through the development of knowledge repositories: Towards work degradation. *New Technology, Work and Employment*, *30*(2), 158–172. DOI: 10.1111/ntwe.12049

Taylor, R. S., Dalal, H. M., & McDonagh, S. T. (2022). The role of cardiac rehabilitation in improving cardiovascular outcomes. *Nature Reviews. Cardiology*, *19*(3), 180–194. DOI: 10.1038/s41569-021-00611-7 PMID: 34531576

Teece, D. J., Pisano, G., & Shuen, A. (1997). Dynamic capabilities and strategic management. *Strategic Management Journal*, *18*(7), 509–533. DOI: 10.1002/(SICI)1097-0266(199708)18:7<509::AID-SMJ882>3.0.CO;2-Z

Teoh, K. W., Ng, C. M., Chong, C. W., Cheong, W. L., Ng, Y. L., Bell, J. S., & Lee, S. W. H. (2024). A digital health-supported and community pharmacy-based lifestyle intervention program for adults with pre-diabetes: A study protocol for a cluster randomised controlled trial [Article]. *BMJ Open*, *14*(10), e083921. DOI: 10.1136/bmjopen-2024-083921 PMID: 39448216

Tesser, C. D., & Norman, A. H. (2021). Quaternary prevention and medicalisation: Inseparable concepts. *Interface: Comunicacao, Saude, Educacao*, *25*, e210101. DOI: 10.1590/interface.210101

Toro-Ramos, T., Michaelides, A., Anton, M., Karim, Z., Kang-Oh, L., Argyrou, C., Loukaidou, E., Charitou, M. M., Sze, W., & Miller, J. D. (2020). Mobile delivery of the diabetes prevention program in people with prediabetes: Randomized controlled trial [Article]. *JMIR mHealth and uHealth*, *8*(7), e17842. Advance online publication. DOI: 10.2196/17842 PMID: 32459631

Trist, E. L., & Bamforth, K. W. (1951). Some social and psychological consequences of the longwall method of coal-getting: An examination of the psychological situation and defences of a work group in relation to the social structure and technological content of the work system. *Human Relations*, *4*(1), 3–38. DOI: 10.1177/001872675100400101

Välimaa, J., Papatsiba, V., & Hoffman, D. M. (2016). Higher Education in Networked Knowledge Societies. In Hoffman, D., & Välimaa, J. (Eds.), *RE-BECOMING UNIVERSITIES? The Changing Academy – The Changing Academic Profession in International Comparative Perspective* (Vol. 15). Springer., DOI: 10.1007/978-94-017-7369-0_2

Vallée, A. (2024). Envisioning the future of personalized medicine: Role and realities of digital twins. *Journal of Medical Internet Research*, *26*, e50204. DOI: 10.2196/50204 PMID: 38739913

Vetrovsky, T., Kral, N., Pfeiferova, M., Kuhnova, J., Novak, J., Wahlich, C., Jaklova, A., Jurkova, K., Janek, M., Omcirk, D., Capek, V., Maes, I., Steffl, M., Ussher, M., Tufano, J. J., Elavsky, S., Van Dyck, D., Cimler, R., Yates, T., & Seifert, B. (2023). mHealth intervention delivered in general practice to increase physical activity and reduce sedentary behaviour of patients with prediabetes and type 2 diabetes (ENERGISED): Rationale and study protocol for a pragmatic randomised controlled trial. *BMC Public Health*, *23*(1), 1–16. DOI: 10.1186/s12889-023-15513-1 PMID: 36997936

von Krogh, G., Nonaka, I., & Rechsteiner, L. (2012). Leadership in organizational knowledge creation. *Journal of Management Studies*, *55*(1), 52–78.

Von Lubitz, D., & Wickramasinghe, N. (2006). Creating germane knowledge in dynamic environments. *International Journal of Innovation and Learning*, *3*(3), 326–347. DOI: 10.1504/IJIL.2006.009226

Von Lubitz, D., & Wickramasinghe, N. (2006a). Networkcentric healthcare: Applying the tools, techniques and strategies of knowledge management to create superior healthcare operations. *International Journal of Electronic Healthcare*, *2*(4), 415–429. DOI: 10.1504/IJEH.2006.010429

Waldman, S. A., & Terzic, A. (2019). Healthcare evolves from reactive to proactive. *Clinical Pharmacology and Therapeutics*, *105*(1), 10–13. DOI: 10.1002/cpt.1295 PMID: 30597564

Wang, L., Tian, W., Zhang, W., Wen, D., Yang, S., Wang, J., Han, X., Wang, J., Ding, W., Wang, L., Yu, Y., & Wu, W. (2024). A Machine learning model for predicting sepsis based on an optimized assay for microbial cell-free DNA sequencing. *Clinica Chimica Acta, 559*, 119716–119716. DOI: 10.1016/j.cca.2024.119716 PMID: 38710402

Wang, R. C., & Wang, Z. (2023). Precision medicine: Disease subtyping and tailored treatment. *Cancers (Basel), 15*(15), 3837. DOI: 10.3390/cancers15153837 PMID: 37568653

Wang, Z., Wang, N., Cao, J., & Ye, X. (2018). The impact of intellectual capital – knowledge management strategy fit on firm performance. *Management Decision, 56*(9), 1865–1884.

Watson, A. R. (2016). Impact of the digital age on transforming healthcare. *Healthcare information management systems: cases, strategies, and solutions,* 219-233.

Weedon-Fekjær, H., Romundstad, P. R., & Vatten, L. J. (2014). Modern mammography screening and breast cancer mortality: Population study. *BMJ (Clinical Research Ed.), 348*(jun17 17), 348. DOI: 10.1136/bmj.g3701 PMID: 24951459

Weick, K. E., & Sutcliffe, K. M. (2001). *Managing the unexpected: Assuring high performance in an age of complexity*. Jossey-Bass.

Weitzel, J. N., Blazer, K. R., MacDonald, D. J., Culver, J. O., & Offit, K. (2011). Genetics, genomics, and cancer risk assessment: State of the art and future directions in the era of personalized medicine. *CA: a Cancer Journal for Clinicians, 61*(5), 327–359. DOI: 10.3322/caac.20128 PMID: 21858794

Wenger, E. (1998). *Communities of practice: Learning, meaning, and identity*. Cambridge University Press. DOI: 10.1017/CBO9780511803932

Wenger, E., McDermott, R., & Snyder, W. M. (2011). *Cultivating communities of practice: A guide to managing knowledge*. Harvard Business Review Press.

Wickramasinghe, N., Haddad, P., Han-Lin, C., Moghimi, H., Delimitros, H. And Vaughn, S. 2015. *Identifying key success factors for the adoption and implementation of CPOE Systems into the Australian private healthcare sector*. Presented at AMCIS 2015.

Wickramasinghe, N. (2006a). Knowledge creation: A meta-framework. *International Journal of Innovation and Learning, 3*(5), 558–573. DOI: 10.1504/IJIL.2006.010489

Wickramasinghe, N. (2006b). Fostering knowledge assets in healthcare with the KMI model. *International Journal of Management and Enterprise Development, 4*(1), 52–65. DOI: 10.1504/IJMED.2007.011455

Wickramasinghe, N., & Davison, G. (2004). Making explicit the implicit knowledge assets in healthcare: The case of multidisciplinary teams in care and cure environments. *Health Care Management Science, 7*(3), 185–195. DOI: 10.1023/B:HCMS.0000039381.02400.49

Wickramasinghe, N., Jayaraman, P. P., Forkan, A. R. M., Ulapane, N., Kaul, R., Vaughan, S., & Zelcer, J. (2021). A vision for leveraging the concept of digital twins to support the provision of personalized cancer care. *IEEE Internet Computing, 26*(5), 17–24. DOI: 10.1109/MIC.2021.3065381

Wickramasinghe, N., & Schaffer, J. (2010). *Realizing Value Driven Patient centric Healthcare through Technology*. IBM Center for The Business of Government.

Wickramasinghe, N., Ulapane, N., Sloane, E. B., & Gehlot, V. (2024). Digital twins for more precise and personalized treatment. In *MEDINFO 2023—The Future Is Accessible* (pp. 229–233). IOS Press. DOI: 10.3233/SHTI230961

Wiig, K. M. (1994). *Knowledge management foundations: Thinking about thinking—how people and organizations create, represent, and use knowledge*. Schema Press.

Wijesekara, P. A. D. S. N., & Gunawardena, S. (2023). A review of blockchain technology in knowledge-defined networking, its application, benefits, and challenges. *Network (Bristol, England), 3*(3), 343–421.

Williamson, S. M., & Prybutok, V. (2024). Balancing privacy and progress: A review of privacy challenges, systemic oversight, and patient perceptions in AI-driven healthcare. *Applied Sciences (Basel, Switzerland), 14*(2), 675. DOI: 10.3390/app14020675

Willis, V. C., Thomas Craig, K. J., Jabbarpour, Y., Scheufele, E. L., Arriaga, Y. E., Ajinkya, M., Rhee, K. B., & Bazemore, A. (2022). Digital Health Interventions to Enhance Prevention in Primary Care: Scoping Review. *JMIR Medical Informatics, 10*(1), e33518–e33518. DOI: 10.2196/33518 PMID: 35060909

Wilson, T. D. (2002). The nonsense of knowledge management. *Information Research, 8*(1), 8–1.

Winstein, C. J., Stein, J., Arena, R., Bates, B., Cherney, L. R., Cramer, S. C., Deruyter, F., Eng, J. J., Fisher, B., Harvey, R. L., Lang, C. E., MacKay-Lyons, M., Ottenbacher, K. J., Pugh, S., Reeves, M. J., Richards, L. G., Stiers, W., & Zorowitz, R. D. (2016). Guidelines for adult stroke rehabilitation and recovery: A guideline for healthcare professionals from the American Heart Association/American Stroke Association. *Stroke*, *47*(6), e98–e169. DOI: 10.1161/STR.0000000000000098

World Health Organization. (2024, December 23). *Noncommunicable diseases*. World Health Organization. https://www.who.int/news-room/fact-sheets/detail/noncommunicable-diseases

Yeo, R. K. (2020). Crossing knowledge boundaries: From team learning to knowledge teams. *Small Group Research*, *51*(6), 700–737. DOI: 10.1177/1046496420919929

Yilmaztürk, N., Kose, İ., & Cece, S. (2023). The effect of digitalization of nursing forms in ICUs on time and cost. *BMC Nursing*, *22*(1), 201. DOI: 10.1186/s12912-023-01333-6 PMID: 37312143

Yu, D. S. F., Li, P. W. C., Li, S. X., Smith, R. D., Yue, S. C. S., & Yan, B. P. (2022). Effectiveness and cost-effectiveness of an empowerment-based self-care education program on health outcomes among patients with heart failure: A randomized clinical trial. *JAMA Network Open*, *5*(4), e225982. DOI: 10.1001/jamanetworkopen.2022.5982 PMID: 35380643

Zaboli, A., Biasi, C., Magnarelli, G., Miori, B., Massar, M., Pfeifer, N., Brigo, F., & Turcato, G. (2025). Arterial Blood Gas Analysis and Clinical Decision-Making in Emergency and Intensive Care Unit Nurses: A Performance Evaluation. *Health Care*.

Zack, M. H. (1999). Developing a knowledge strategy. *California Management Review*, *41*(3), 125–145. DOI: 10.2307/41166000

Zanotto, B. S., da Silva Etges, A. P. B., Marcolino, M. A. Z., & Polanczyk, C. A. (2021). Value-based healthcare initiatives in practice: A systematic review. *Journal of Healthcare Management*, *66*(5), 340–365. DOI: 10.1097/JHM-D-20-00283 PMID: 34192716

Zeng, M. L., Hong, Y., Clunis, J., He, S., & Coladangelo, L. (2020). Implications of knowledge organization systems for health information exchange and communication during the COVID-19 pandemic. *Data and Information Management*, *4*(3), 148–170. DOI: 10.2478/dim-2020-0009 PMID: 35382097

Zhou, H., Liu, F., Gu, B., Zou, X., Huang, J., Wu, J., Li, Y., Chen, S. S., Zhou, P., & Liu, J. (2023). A survey of large language models in medicine: Progress, application, and challenge. arXiv preprint arXiv:2311.05112.

Zipperer, L. (2016). Knowledge Management in Healthcare. Routledge. https://books.google.com/books/about/Knowledge_Management_in_Healthcare.html?id=gmUGDAAAQBAJ

About the Authors

Nilmini Wickramasinghe is the inaugural Optus Chair and Professor of Digital Health at La Trobe University. She is also an inaugural member of the Australian Centre for AI in Medical Innovation. She holds honorary research professor positions at the Murdoch Research Children's Institute (MCRI), the Peter MacCallum Cancer Centre and Northern Health. After completing 5 degrees at the University of Melbourne, she was awarded a full scholarship to complete PhD studies at Case Western Reserve University, Cleveland, OH, USA and later she was sponsored to complete executive education at Harvard Business School, Harvard University, Cambridge, MA, USA in Value-based HealthCare. For over 20 years, Professor Wickramasinghe has been actively, researching and teaching within the health informatics/digital health domain in US, Germany and Australia with a particular focus on designing, developing and deploying suitable models, strategies and techniques grounded in various management principles to facilitate the implementation and adoption of technology solutions to effect superior, value-based patient centric care delivery. Professor Wickramasinghe collaborates with leading scholars at various premier healthcare organizations and universities throughout Australasia, US and Europe and is well published with more than 400 referred scholarly articles, more than 30 books, numerous book chapters, an encyclopaedia and a well established funded research track record securing over $25M in funding from grants in US, Australia, Germany and China as a chief investigator. She holds a patent around analytics solution for managing healthcare data and is the editor-in-chief of two scholarly journals published by InderScience: Intl. J. Biomedical engineering and Technology and Intl. J Networking and virtual Organisations as well as the editor of the Springer book series Healthcare Delivery in the Information Age. In 2020 she was awarded the prestigious Alexander von Humboldt award for outstanding contribution to Digital Health, the first time this honour has been bestowed to someone in the discipline of Digital Health

Nalika Ulapane (PhD, BSc (Engineering)) is a researcher specializing in mathematical modelling, engineering systems, machine learning, and digital health. He holds a PhD in Engineering and a BSc in Electrical and Electronics Engineering. Dr. Ulapane has held various research and teaching positions at prestigious institutions, including the University of Melbourne, La Trobe University, Swinburne University of Technology, and the University of Technology Sydney. Dr. Ulapane solves complex problems in engineering and information systems, healthcare, and artificial intelligence by combining mathematical modelling, engineering systems design, data analytics, machine learning, and design science research principles. He has made notable contributions to electromagnetic sensing, robotics, and digital twins for healthcare. Dr. Ulapane has also served as a guest editor for the Sensors Journal (MDPI) and a book editor for several publishers. He has co-authored numerous publications, including award-winning and well-cited articles. Currently, he is a Research Fellow at La Trobe University.

Ayesha Thanthrige is a PhD candidate at the School of Computing, Engineering, and Mathematical Sciences at La Trobe University, Australia. With over a decade of experience as a Business Analyst and Project Manager in the field of information technology and as a certified Scrum Master, Ayesha specializes in leveraging digital technologies to drive innovation in healthcare. Their research, "Digital Transformation in Healthcare: The Case of Personalised Self-Management for Wellness and Healthier Lifestyle Understanding," focuses on designing intelligent, patient-centered systems that promote sustainable healthcare solutions. Ayesha holds a BSc (Hons) in Management Information Systems from University College Dublin and a Master of Business Administration from the University of Sri Jayewardenepura. Ayesha has co-authored several publications in the digital health domain and actively contribute as a reviewer for international conferences. Passionate about responsible and intelligent healthcare innovation, Ayesha brings valuable expertise in digital transformation and interdisciplinary collaboration to this project.

Index

A

B

C

D

E

F

G

H

I

K

M

O

P

Q

R

S

T

www.ingramcontent.com/pod-product-compliance
Lightning Source LLC
LaVergne TN
LVHW081521300926
849161LV00006B/1254